BE YOUR OWN BUILDER

FOREMAN

BE YOUR OWN BUILDER

HOW TO DESIGN AND BUILD YOUR OWN HOME

Keith
'The Blockinator'

Published in 2024 by New Holland Publishers
First published in 2015 by New Holland Publishers
Sydney

Level 1, 178 Fox Valley Road, Wahroonga, NSW 2076, Australia

newhollandpublishers.com

A catalogue record of this book is available at the National Library of Australia.

ISBN: 9781760797645

Managing Director: Fiona Schultz
Publisher: Diane Ward
Project Editor: Susie Stevens
Design: Andrew Davies
Production Director: Arlene Gippert
Printed in China

Keep up with New Holland Publishers:
NewHollandPublishers
@newhollandpublishers

This book is dedicated to my amazing mother Judy Schleiger, who always stood behind me. Thanks Mum – rest in peace.

Keith Schleiger is a carpenter, builder and foreman with over 27 years of experience in the building industry. As a highly qualified expert in his trade, Keith has been involved in building just about every type of structure there is, from schools and hotels to multi-storey office buildings and million-dollar mansions.

In 2011, Keith joined the hugely popular reality TV show *The Block*. Known as 'The Blockinator', his knowledgeable expertise coupled with his keen eye for detail and easy-going nature quickly made him a popular addition to the team. The past six seasons of the show have seen Keith guide dozens of amateur builders from every walk of life through the process of renovation.

In addition to his work on *The Block*, Keith recently finished building a house for himself and his family in the southeast suburbs of Melbourne – and learning a lot about the building process from the perspective of a homeowner.

CONTENTS

ABOUT THIS BOOK

This book is meant to serve as a user-friendly guide, so here are a few things to look out for.

Building is a specialised field with specialised terms that may be unfamiliar to the amateur builder. When you come across a **technical term** in bold, you should see its definition nearby, and if you forget what something means, there's a **glossary** of all the terms I use at the end of the book. Also keep an eye out for the following helpful hints.

Q&A

I've included plenty of other hints and suggestions to help answer your questions and see your build go smoothly. You'll find these under the Q&A heading.

I've gained a lot of experience throughout my years as a builder and from my experience of building my own house. Look for these words to the wise as you go.

CHAPTER SUMMARY

Every chapter ends with a quick recap of the major points we've just covered in case you need to refer back.

WHAT'S NEXT

The last words in each chapter let you know what's coming up in the next chapter.

INDEX

We've included an index of terms and the pages that they appear on so you can quickly find where that subject matter appears in the book.

FOREMAN

INTRODUCTION

ME? BUILD A HOUSE? NOT A CHANCE!

For someone who's not a professional tradesman, the idea of building your own house might seem a little overwhelming. And while building a house from the ground up is a big job, you might be surprised how easy it actually is to be your own builder.

POLLY AND WAZ – THE LEAST QUALIFIED BUILDERS ON THE BLOCK

When I first met Polly and Waz, I reckoned their chances for success on *The Block* were pretty slim. Billed as the city girl and the country boy, they were about as 'unhandy' a couple as I'd ever seen. Waz might be a country bloke, but as a graphic design student, he didn't know the first thing about building. And Polly, the daughter of one of Australia's most well known doctors, knew even less.

Like the rest of the contestants that season, Polly and Waz started with nothing more than the perimeter of a house. A professional contractor put in the subfloor and the roof of the structure for them, but building the rest of the three-bedroom/two-bathroom home was up to these two amateurs, the least experienced of all the teams that year. They were the underdogs from the beginning.

But what Polly and Waz lacked in know-how, they made up for in hard work. They had to design everything from wall colours and floor panelling to window dressings and doors - the lot - so Polly spent hours and hours reading through magazines and researching online looking for ideas. Waz worked just as hard, throwing himself into the dirty work and learning a lot in the process.

One thing they didn't really do, though, was the majority of the manual labour. Instead, they brought in electricians, plumbers, carpenters and other qualified tradesmen to carry out the sort of specialty work most people can't - and shouldn't - do themselves.

In my role as the **foreman** of *The Block*, it's my job to oversee every aspect of the projects the contestants are working on and make sure they're building to code. Every nail, every screw, every piece of timber has to be perfect because at each stage, an independent building surveyor will come in and look over the work. If the contestants don't pass inspection, they have to do the job again until it's right. It can be the difference between success and failure.

A **foreman** is in charge of overseeing your build and liaising between you and the rest of your employees.

A lot of work goes into building a house and, trust me, I should know. I've been a builder for over twenty years. I've seen first hand all the details that go into taking a project from an idea to a reality. Most of the people who come on *The Block* haven't – including Polly and Waz. More than once, I held my breath and hoped they would get it right.

As the season progressed, I watched this unlikely couple build their house from the foundations up. It wasn't easy. They were on a very tight budget – the contestants are normally allotted about $12,000 per room, a figure that includes all of the building materials and labour plus all of the furnishings as well. However, they were able to stick to their budget by doing deals with the suppliers and doing a lot of the work themselves.

Despite their lack of experience, Polly and Waz did have a few things going for them. They worked together as a team really well and they were both flexible and willing to learn – in the end, their persistence and hard work paid off. They ended up winning the season and walking away with over $100,000 for their efforts.

WHY NOT BUILD YOUR OWN HOUSE?

For someone who's not a professional tradesman, the idea of building your own house might seem a little overwhelming. Buying a house is the biggest investment most people will ever make and if you decide to build, you want the job to be done properly - perfectly, in fact. And while building a house from the ground up is a big job, you might be surprised how easy it actually is to be your own builder.

I recently went through the process myself when I decided to oversee the design and construction of my own house. I know I'm a professional foreman, so right now you might be thinking, 'Of course Keith the Blockinator can build a house - that's what he does for a living!' And you'd have a point.

I have to agree that I'm more than qualified to deal with just about anything that comes up on *The Block*, but even with all of my experience, I still learned a lot when it came to building for myself and my family, especially in terms of finance and logistics. And even though I know how the jobs should be done, I hired most of the tradesmen who did the work. I'm a qualified carpenter so I was able to do all the carpentry work myself, but I had all of the plumbing, electrical, plastering and other specialist trades done by contractors I hired.

My point is that you don't have to be a professional builder or tradesman to build a house. If you know what you're doing, know the right questions to ask and the right people to employ, you really can do it yourself. With just a little bit of know how, you can take on the role of owner builder and enjoy a huge amount of control over the design and construction of your home while saving a heap of money at the same time.

Get in there

Don't be afraid of the job site. A starting apprentice doesn't have any training or experience at all. Most apprenticeships include hands-on training plus coursework towards a certificate, but the two things are usually happening at the same time. Apprentices will work on the job during the week and attend classes a couple of evenings a week or on the weekends.

As an owner builder, you don't have any less experience than a new apprentice. You should be able to follow directions at least as well as a newly-graduated high school student. At the same time, you also need to know when to stop. A lot of jobs are very repetitious and don't require a huge amount of skill, but as soon as you don't feel confident doing a job, hand it over to someone else.

As long as you know where to draw the line, there's no reason for you to pay an unskilled teenager $40 an hour to do simple labouring work when he or she doesn't have any more skills than you have.

WHY SHOULD I LISTEN TO YOU, KEITH?

That's an excellent question. Throughout this book, I'm going to tell you how to go about hiring other professionals. One rule of thumb is that everyone you hire should be able to demonstrate his or her qualifications and experience – and that rule applies just as much to me.

I first became a builder sort of by accident. When I completed my VCE, Victorian Certificate of Education, I had intended to be a schoolteacher, but the summer after I graduated from high school, I took a job working with a builder to make some extra money. I really enjoyed the work so I decided to pursue an apprenticeship in carpentry.

From the age of eighteen, I worked doing office fit outs. My boss at the time wanted to continue on with offices, but that wasn't really what I wanted to do, so I asked him for an early release. Normally an apprenticeship takes four years, but I was able to go out on my own as an independent tradesman after only three.

Immediately after ending my apprenticeship, I started working with a friend who specialised in building high-end houses. Over the course of the next year, I was able to learn every aspect of that industry including general house framing, formwork concreting (the type you see in high-rise buildings), how to build an underground car park – everything.

Throughout that year, I learned how to do really unique forms of carpentry, the type you don't see every day. These were houses that were worth $2 million or more so we had to get it right. It was a great training ground in which to learn all about attention to detail and staying on schedule.

After twelve months of working with my friend and mastering my trade, I took on a contract of my own. I was 21 years old. I did all the carpentry to build an entire primary school and from there my career really blossomed.

After that, I did more high-end housing, doing carpentry contracts with partners. I also worked on large commercial projects doing apartments. I became very efficient and

I had learned the hardest parts of building at a young age. By 24, I could basically build anything.

At 26, I took on my first job as site foreman. I converted a large section of St Vincent's Hospital in the Melbourne suburb of Fitzroy into apartments which went on to sell for a lot of money. From there, I moved to a project converting a 150-bed hotel into a 400-bed hotel in St Kilda – that project took three years.

Over the next twenty years, I built every type of structure imaginable, staying generally in the high-end housing sector of the building industry. I was able to make a good living and, more importantly, I loved it. I really enjoyed the challenge. I've built houses in Armidale, Toorak and Brighton – some of Melbourne's most exclusive neighbourhoods. I've done jobs where I supervised ten or fifteen people at once and jobs where I supervised 150 people daily.

For the last four years, I've been the site foreman on *The Block,* a position I lucked into through word of mouth and being in the right place at the right time. I had been working for one of Melbourne's most elite high-end builders when one of my colleagues, a project manager who was also a mate, went to work for a different company. When his new company was appointed to be the official builder for *The Block*, my mate rang me up and said, 'You'd be a great fit for us on *The Block* as a foreman.'

Since then, I've overseen every aspect of every build for six seasons worth of home renovations. It hasn't always been easy helping amateur builders stay on track, but I'm really pleased with the hard work that all the contestants have put in over the years, and I'm glad to have been a part of it.

A Word to the Wise

Watch your back

A job site can be a great place to work because we're always joking around and having a laugh while we work. But be warned – lots of practical jokes happen on the job site, even on *The Block*. In the Glasshouse season, brothers Simon and Shannon played lots of jokes on me. They kept locking me out of my office, and I also caught them dropping wet concrete on my head and trying to impersonate me. Finally, I reckoned I'd get them back.

One night while they were out, I sneaked into their apartment. I knew they were having a big night and that they wouldn't be back till very late, so I mixed up a couple of buckets of concrete for them. Around three or four o'clock in the morning, they came stumbling in, pulled back their doona and hopped right into a nice soft layer of wet concrete. I wasn't completely heartless, though – at least I covered up their bed with plastic before I started pouring!

IF THEY CAN DO IT, SO CAN YOU

You might be looking at building a brand new house, or it could be a better idea for you to knock down an old building and start over. Maybe you just want to do some renovations to your existing home. Whatever you want to build, this book aims to guide you through the process including why you should consider being an owner builder, how to get started, who to hire and what to expect at each step along the way.

Using the expertise I've gained on the job, my experiences in building my own house and the extensive amount of research I've done, I can tell you what you need to do and how you need to do it to see your ideas come to life. Not sure what you want? That's okay, too – I can help you decide.

In my role as the foreman on *The Block* I've seen six blocks of apartments successfully renovated by complete amateurs. I've worked with people from all walks of life, many of whom have never touched a tool in their lives – from chimney sweeps to concert promoters, receptionists to police officers.

None of these people had any building experience. What they did have, though, was management skills, the right attitude and a passion for the job. And with only those few qualities, they were able to build some fantastic homes. If they can do it, anybody can. Why not you?

CHAPTER ONE

WHY SHOULD I GIVE IT A TRY?

These days, more and more people are choosing to build their homes rather than buy an established one for a lot of good reasons. There's an old saying that goes 'they don't build 'em like they used to' and I have to agree – when it comes to homes, we're building way better than we used to.

The construction that's available today is more advanced than ever. Building techniques and materials are improving all the time through advancements in science and engineering which means that a brand new house built today is likely to be more sturdy, more energy efficient and more personal than those built even five or ten years ago.

Even if you don't want to build a brand new house, it's worth considering rebuilding an existing house or updating your current one to make it as energy efficient as possible while maximising the amount of space you have. (Don't think you have room to grow? Consider going up - we'll talk more about the possibility of adding a second storey in Chapter Two.)

WHO SHOULD BUILD YOUR HOUSE?

If you decide that building a house is right for you - either on a new block of land or to replace or renovate an established home - you have a couple of options in terms of builders.

Volume Builders

One option is to go through a large volume builder. Using a large company is tempting because they're easily accessible and they usually deliver on time. You can also go and see their range of display homes, which can give you a real feel for what the finished product will look like before you start. Going big is definitely an easy route to take.

But quite often, the big guys will use the same **contractors**, and those contractors are always busy - sometimes too busy. When tradesmen are under a lot of pressure to get a job done quickly, they may opt for speed over quality. Things may get left out or overlooked in the rush. You might end up with bowed walls, floors and ceilings out of level, walls not straight or any number of other imperfections that you will have to either live with or pay to have fixed.

Contractors – also called tradesmen – are the skilled workers who provide the labour and/or materials necessary to build your house. The ones you're most likely to need are concreters, carpenters, plumbers, electricians, plasterers and painters.

A **premium** – also called a margin – is the amount of money a builder will charge you in addition to the cost of labour and materials. Basically, it's the amount you pay that the builder takes as profit.

Also, most builders will charge you a **premium** of up to thirty per cent to build your house. That's not for doing the framing or bricklaying - that's simply to organise and oversee the tradesmen who will do the actual hands-on work.

And depending on the size and shape of your land, the volume builders may not have a style that will work for you. If you look at the options on offer, you'll quickly see that the designs of most volume builders are based on the length and width of your property. If you have a smaller block or a plot of land that is an abnormal shape, their designs may not physically fit within the boundaries of your property.

Finally, if you want to demolish an old house and rebuild or you plan to do renovations, the volume builders probably can't help you.

A Word to the Wise

Know your costs

If you hire someone else to build your house, you'll never really know what's going on with the project. I once knew another builder who was renovating a very expensive home. The designs included a crucial outdoor structure, but this builder didn't start working on that structure until after the foundations and the walls of the house were up.

They were able to include that outdoor structure, but it cost an additional $40,000 to add it so late in the project - all of which was paid by the clients who never knew the builder had made the mistake. When you're the owner builder, you know everything that's happening with your house.

Independent Builders

Another option is to find a smaller independent builder. Any contractor working for a large building company will almost certainly be earning less money than if that person were working independently.

In contrast, the small builder has to have a good reputation and provide quality work to be able to remain independent of the big building companies. Independents are obviously good at what they do if they can stand on their own two feet and survive on their own ability.

At the same time, every builder will charge you a margin, and fair enough – that's how builders make their money. An independent builder may charge you more to build your home because smaller builders probably can't buy the materials in bulk the way a volume builder can. Volume builders can afford to buy in massive quantities and they usually get a discount when they do. Independent builders just won't need the same quantities and they will need to make up the difference.

Another factor to consider is the financial position of an independent builder. Volume builders can usually weather a loss, but a smaller builder sometimes can't. What happens when something goes wrong – even if the problem has nothing to do with your build – and your builder can't afford to stay in business? If your builder goes broke halfway through your project, the person left holding the bag is you.

And just like with volume builders, an independent builder will still have the final say in what happens on the project. Instead of you, he'll be the one to hire all of the tradesmen and he'll be the one to tell you if those contractors are producing quality work. While an independent builder may be more caring and attentive to your preferences than a volume builder, it is your builder who still holds the majority of control.

Do It Yourself

Perhaps the best option – the one that a lot of people don't have the confidence to consider – is to oversee the construction of your house yourself as an **owner builder**.

What does it mean to be an owner builder? The short answer is that you are the project manager for your build.

An **owner builder** is someone who takes on most of the responsibility for domestic building work carried out on his or her land.

You won't be the person swinging the hammers or digging the foundation; instead, you'll be responsible for hiring the right people to do the majority of the work for you. You'll also be accountable for making sure the work is done properly so that it passes inspection and meets all the building codes required by law in your state.

It may sound a little scary, especially if you don't think you're qualified to determine how well jobs have been done, but you don't need to have all the expertise - you just need to hire the right people who do. As an owner builder, you will be the person in charge, and there are lots of benefits to taking on that role.

Savings

Perhaps the most attractive reason for doing it yourself is the amount of money that stays in your own pocket. Even a small builder will want to make a profit for organising the job, but unlike jobs such as plumbing or electrical work which require a lot of expertise to do well, there's no special training necessary to oversee the project.

More house for the money

With the new building materials that keep coming in, it's now possible to build a high-end product for a cheaper price. Also, high-end builders (both volume and independent) will charge bigger margins - their profits are bigger than those of builders who produce smaller,

less expensive homes. Instead of paying an extra twenty or thirty per cent in profit to someone else, that money can go into the construction of your house.

Personal preference

As an owner builder, you get to oversee the design of your home from the very beginning. Rather than ending up with a stock standard Australian house that looks the same as every other house on the street, you get to create exactly what you want. Instead of being limited to the handful of options a volume builder will provide, you can look at thousands of options from different sources and then cherry pick the bits you like best from each (we'll talk more about designing your house in Chapter Three).

Security

When you're the one in charge, you get to hire every tradesperson who comes to work on your house, which means you have a lot more control over who is coming in and out of your property. You also get to meet each tradesman on a personal level. Let the big builders do it and you'll probably never meet any of the tradesmen.

Control

If you're building yourself and you decide you aren't happy with the work that's being done, you can always end a contract. You're in control all of the time, which means that you can (usually) change your plans, your preferences and your contractors whenever you want to. You're also in charge of quality control so you can make sure any small problems or flaws get fixed quickly before they turn into bigger problems that are more difficult to amend.

Ownership

The sense of accomplishment that comes with buying a house is one thing, but knowing that you were behind the construction of the entire project is even better. By doing it yourself, you can feel confident that you got what you wanted and you can enjoy the sense of pride that comes from being directly involved in a job well done.

Longevity

It's no secret that, whether you buy an established home or build, moving house is expensive. Instead of buying two or three homes that aren't quite what you want, it might cost you less money in the long run to build exactly what you want the first time and stay in the same house for a longer period of time.

Flexibility at a lower cost

As an owner builder, you'll have the ability to make changes at a cheaper cost than if you were using a contracted builder. Quite often changes do occur in building. Clients will get halfway through and decide to make a room bigger or turn a window into a door - and that's when builders make their money. They will charge you like a wounded bull to make those variations, as much as $100 an hour even for something simple. If you're doing it yourself, you can usually make these changes with minimal cost.

A Word to the Wise

Enjoy the savings

There were a lot of factors that prompted me to take on the project of building my own house, but I can safely say that (at least in the beginning) savings was probably the biggest motivator for me. I did encounter a lot of problems to do with financing (I'll talk more about those issues in Chapter Four), but in spite of all of my problems, I reckon I've saved at least $200,000 in the end by doing it myself.

THE PITFALLS OF DOING IT YOURSELF

It would be unrealistic for me to suggest that there's no danger involved in building your own house. The fact is that, if you do something wrong, you could be very sorry. Before you make your final decision, there are a few risks to consider.

Legal issues

Being an owner builder means you have more control over your build, but with that control comes a significant amount of responsibility. In Victoria, for example, the current law states that an owner builder must live in the house they build for at least three years before they sell, but they must be able to guarantee the workmanship that went into the house for a minimum of six and a half years. If you hire a builder, that responsibility goes with them. If you build the house yourself, the responsibility lies with you.

Also, if you don't have the proper type and amount of insurance, you will be held liable if someone is injured on your property during the build. We'll talk more about what insurance you need in Chapter Four and how to create a safe working environment in Chapters Eight and Nine.

Safety

While I would encourage you to hire qualified tradesmen to do the actual work of the build, there will always be jobs you can do around the job site. Having said that, if you don't take proper care about where you go, what you wear and how you work, you can hurt yourself or someone else very badly. For this reason, it is paramount that you listen to the advice of the qualified experts on site and take all necessary precautions. We'll talk more about how you can work safely in Chapters Eight, Nine and Ten.

Time

You don't need to be on the job site every minute of every day, and there will be times when the best thing you can do is to leave. In saying this, the job of building a house is a big one and it will require your attention. Legally, you can leave all the actual hands-on work to someone else, but you do need to be supervising in some capacity, even if that's via a foreman. We'll talk more about how to hire one in Chapter Six. If you have zero time to devote to the project, though, you probably shouldn't take on a commitment of this size.

Build quality

If you've seen much of *The Block*, you'll notice that the contestants are constantly bringing in qualified tradesmen.

There's a good reason they hire contractors who have the proper training and experience to do specialised work. If you try to do too much yourself, or you hire people who aren't very good at their jobs, you may end up with a substandard product. Even if you do manage to pass your inspections, who wants to live with a house full of flaws?

Stress

There's no doubt that building your own house will be a stressful experience. If you anticipate every aspect of your project to go smoothly, you should adjust your expectations now. No matter how well prepared you are, you will surely experience a lot of anxiety in getting the job done properly. I've heard it said, though, that there is little in life more rewarding than working hard at work worth doing – what could be more rewarding than building your own house?

A Word to the Wise

Be safe

Safety on the job site is really no joke. One time, I was working with another carpenter who was using a power saw. Most hand-held power saws come with a fence, which is the safety guard, and as soon as you stop cutting, the fence comes down. Well, my fellow carpenter reckoned the fence was getting in his way and that he could work faster without it, so he took it off.

In the middle of the job, he stopped to talk to someone, not thinking that the fence was off the saw. Without thinking, he let the saw fall to his side, and he sank the blade right into his thigh. He had a gash of about 120cm long and two or three inches deep that cut through all of his nerves and muscles.

After cutting himself, that carpenter had to get a lot of stitches and it took him a long time to heal, but it could have been worse. He could have got himself in between the legs or in the stomach or somewhere worse.

You don't want to hurt yourself. Aside from the pain, a serious injury can take you away from your regular job and, even if you have income protection, it can really set you back. In other words, be careful on the job site and always have safety in the front of your mind.

FINANCING AN OWNER BUILD

If you want to build your own house, I cannot stress enough how important it is to get your financing right. If you miscalculate the amount of money you need to borrow to get your job done, you may find yourself in real trouble. You should know up front that getting finance as an owner builder is very different than getting financing to buy an established house or to have someone else build your house.

When you buy an established house, you can usually start with a small down payment and borrow the rest. As an owner builder, you'll need to be prepared to come up with a hefty portion of the construction costs up front or from funds not covered by your loan.

If you're a **credit risk,** you probably won't be able to get any financing to be an owner builder at all.

A **credit risk** is someone who has failed to pay his or her bills on time in the past. A lender won't want to loan you money if you're a credit risk because they have reason to believe – based on your previous borrowing history – that you may not pay them back.

We'll talk more about organising your finance in Chapter Four and about ideas for drawing up a realistic budget in Chapter Seven.

SETTING A REALISTIC TIME FRAME

If you're a fan of home renovation shows like *The Block*, there's every possibility that you've developed an unrealistic expectation of how much time a particular project should take. The contestants on the show are able to complete an extensive home renovation in ten weeks, but in reality that would probably never happen. The kitchen that takes a week to complete on *The Block* might take a couple of months to build in the real world.

Why? For one, a lot of the groundwork is done for the contestants, things like demolition and obtaining all the necessary reports and licenses. Also, on the show, we're working around the very rigorous time constraints of television production, which means contractors will sometimes bring in two or three times as many members of their crew for a project on *The Block* as they would for a typical job.

Exactly how much time you should expect to allow for your building project will depend on a lot of factors including what you want, the dimensions of your property, the type of soil you have and local council regulations. You should be able to determine a fairly accurate assessment of how much time to allow for your build before you ever start. We'll talk more about drawing up a master schedule in Chapter Seven.

WHO SHOULD BE AN OWNER BUILDER? WHO SHOULDN'T?

By now, you may be thinking that being an owner builder would be a great option for you. If you're patient, flexible, attentive and willing to work hard, you're likely to be a fantastic owner builder. You'll also do well if you have a keen eye for detail and you're a good listener. It is a big job, though, so before you commit to building your own house, there are a few questions you need to ask yourself.

I don't have any building experience at all – can I do it myself?

Actually, you don't need any experience in building or trades. As the owner builder, your job will be to organise experts in their fields to do the hands-on work for you. The fact is that there's a lot you probably can't do yourself and shouldn't try - we'll talk more about why you want to hire qualified tradesmen in Chapter Seven - and sometimes the best thing you'll be able to do it just get out of the way.

Having said that, there will still be jobs you can do around the job site yourself, no matter how unqualified you may be, and doing those jobs will allow you to be more involved and will save you money. We'll cover those tasks in the sections dealing with the actual build in Chapters Eight, Nine and Ten.

My partner and I both work full time – will we have enough time?

In all honesty, the more time you can spend on the job site, the better. If you have zero time to commit to building a house then you probably shouldn't consider taking on the project. However, you don't need to commit to being there all the time and, with a small renovation, you may not need to attend at all.

If you're building a bigger project with your partner, you may be able to arrange for one of you to visit the job site in the mornings for half an hour before work a few days a week and the other to visit on the way home from work. You can also turn up whenever you want, and it's not a bad idea to plan on dropping by unannounced on occasion, just to make sure things are happening in your absence.

It would be ideal if you (or your partner if you have one) could take a bit of time off from work or use some holiday time, even if it's only for part of the project. The good news is that between smart phones and email, you've always got ways of keeping in touch with your foreman. We'll talk more about this vital member of your team in Chapter Six.

Even if you can't be there every day, you should really commit to overseeing a site meeting for at least an hour every week of the build. Always remember, it's your home. It's a huge investment so be all over it as much as you can and you'll definitely benefit.

A Word to the Wise

Make the time

My wife and I were able to manage fitting our build into our busy schedules, but it was hard. I had to spend a lot of time filming while my house was being built. At the same time I was able to oversee my project mainly by making and taking several phone calls a day.

My wife was also able to manage a lot of the paperwork via the Internet and she was constantly sending emails. Together, we spent many hours a day approving quotes and looking at designs. But it only takes a ten-minute phone call every day to your foreman to keep the project moving along.

Sometimes you'll need to answer a question or deal with a problem immediately, but a good foreman should be able to keep the project on track and discuss a list of the day's issues with you after hours. Before you begin, you should work out with him or her the best way of staying in touch.

I'm not sure I have the right personality to pull off this project

While most people are probably capable of acting as an owner builder, some people will be more suited to the job than others, and certain personalities may find it very difficult.

The timid mouse

If you're the sort of person who doesn't know how to ask for what you want, you're probably not going to end up with the house you had in mind. Act wishy washy or uncertain and your tradesmen are going to have to guess what it is you're after.

Also, some people find tradesmen daunting, but the building industry has definitely improved as far as tradesmen's attitudes towards clients are concerned. Even if you aren't the most assertive person, you should expect your contractors to be professional. They have to be because how they provide their services to their clients is a big part of their business. I think tradesmen have come a long way over the last ten or fifteen years in terms of their manner, so don't be intimidated. If someone is rude or not willing to work with you, consider not hiring them or, if the person is already on the team, you can fire him or her – that's part of the reason you're the owner builder.

The hot head

Another person who won't do well as an owner builder is someone with a bad temper. You have to be able to remain calm under stress because you will definitely encounter stress along the way.

On *The Block*, I sometimes have a reputation for being a bit of a hard ass, but there's a difference between being firm and being angry. The fact is that it rarely helps to lose your temper – which is why I usually don't. If you're trying to manage a team of tradesmen, a boiling temper can really work against you.

It's a good idea to be honest and speak up if you think something hasn't been done properly, but if you don't know how to manage your anger, being an owner builder might not be a good idea for you. The person you hire to act as your foreman, however, can (and probably should) liaise on your behalf with the tradesmen. As long as you can maintain a good working relationship with your foreman, you may be able to avoid situations that are most likely to turn confrontational.

The waffler

If you can't make decisions, managing a building project will be especially hard for you. It's always important to do your homework but, at many stages along the way, you'll need to be able to make a choice. Make no mistake, you'll have a lot of big decisions to make and if you take too long, it can hold up the job.

There are times when you'll have to allocate two or three or four hours at night to make decisions. If you take your time with the planning stage (and you should), you will have made most of those really big decisions long before the actual construction begins.

The know it all

If you're the type of person who thinks you can wing it and do things the way you think they should be done rather than according to regulations, you should maybe reconsider being an owner builder.

When it comes to construction, even a simple renovation, if you don't abide by the law, you won't win. You won't be able to simply pay off a **building inspector** or a council. They'll make you pull the house down if you haven't followed the rules. You can't be arrogant and think, 'I'm not going to worry about that.'

The **building inspector** – sometimes called the building surveyor – will check your project at various stages throughout the build to ensure that all of the work has been done properly. When you pass the final check, the inspector will give you a Certificate of Occupancy.

Likewise, if you want to help and you don't listen to the advice of the professional tradesmen on site, you are very likely - almost certain, in fact - to make a huge mistake. You'll have to live with the consequences of your own workmanship, and when you try to sell your home, you'll find the value of your house won't be as high as it could have been.

A Word to the Wise

Manage your people

One skill you really must have as an owner builder is people management skills. You need to have a professional relationship with your employees and get across what you want without being a push over or being impolite.

Here's an example: your bathroom needs to be waterproofed, and your tiler is the one to give you that certification - no one else is required to inspect that job. Now, let's say you're really rude to your tiler and you alienate him or her. You might find that your waterproofing is substandard or that it hasn't been done at all.

It would be very unprofessional and I don't think very many tradesmen would do it, but it's something to think about. You'd never know until years later when the water damage is so extensive that you have to replace the entire bathroom.

Act like a jerk to your plasterer and he or she might not thoroughly screw up all of your cement sheets. You won't get the best result and you won't know until four or five years later when all of your walls start to crack.

You have to have the humility to listen to the experts and abide by the rules, regardless of what you actually want. If you don't, you'll probably make a terrible owner builder.

At the same time, you don't want your tradesmen to think they can show up late, take off early or take their time getting their jobs done. Sticking to your schedule is of vital importance, so you need to make sure your contractors take you seriously. We'll talk more about staying on track in Chapters Seven and Eight.

If you can be firm but nice at the same time, you'll get tradesmen who want to go the extra mile for you, and maybe stay back a little while to get the job done or do a bit more for you than what's required of them. If a tradesman doesn't like you, you run the risk of getting the bare necessity done rather than an excellent job. The tradie might resent you and you won't know about it until five years down the track when your house starts to break down.

A Word to the Wise

Don't hire friends or family

If you have a very close friend or family member who is a tradesman, you may want to reconsider becoming an owner builder. It's generally not a good idea to hire people who are close to you for building your own house – if you know someone who is going to be angry with you for choosing someone else, you might want to think again. We'll talk more about hiring the best tradesmen and getting the most out of them in Chapter Seven.

If you want to be an owner builder, there are a lot of factors to consider, but keep in mind that heaps of people with as little (or less) experience than you have built really amazing homes and created a lot of equity for themselves. Make no mistake, building a house is a big job but if you do it right, it will be an investment that can last a lifetime.

CHAPTER SUMMARY

- Because of advances in building technology and materials, now is a really good time to consider building, rebuilding or updating your house.

- If you want to build, you have a few options – you can sign on with a volume builder, hire an independent builder or do the job yourself.

- There are a lot of benefits of being an owner builder. These benefits include getting more house for the money, having more room to include your personal preferences, security, control, ownership, longevity and flexibility.

- There are definitely risks involved in being an owner builder. These risks include legal issues, safety, time, build quality and stress.

- Make sure you will be able to finance your project. You won't be able to borrow the entire amount that you anticipate needing, so make sure you have some funds beyond your loan. If you are a credit risk, you probably won't be able to finance as an owner builder at all.

- Remember that television isn't the same as real life. Everything happens a lot faster on *The Block* than you can expect to happen in a real-world situation, so be prepared to keep your expectations realistic.

- Certain types of people are better suited to being owner builders than others. You will need to have some time to devote to the project. If you have no time at all, it will be extremely difficult.

A few personality types should think long and hard before taking on a project of this scale, for example, people who can't speak up for themselves, people who can't control their temper, people who can't make decisions and people who think they know everything. Don't worry if you think you don't have enough experience – you don't need to.

It's rarely a good idea to hire friends or family to build your house. It's better to let business be business and for you to remain friendly but impersonal with your contractors.

WHAT COMES NEXT

By now, you've thought a lot about whether or not you'd be capable of being an owner builder, whether or not you can afford it and whether or not you're suited to the job. If you think yes, now it's time to get stuck into the design stage.

CHAPTER TWO

GETTING STARTED

Now that you've decided to become an owner builder, you may be eager to start building – slow down! What comes next is the research phase, and it's essential that you take your time with it. Planning and designing your project should probably take longer than the rest of the build combined.

A Word to the Wise

Check the details for your situation

Please keep in mind that the information I've set forth in this book is meant to be taken as a guideline. There are many places along the way when what you need to do may be different to the examples I've given. Why? Every job will be different and rules vary from one state or council to the next. I'm here to give you an overview, but you'll need to confirm exactly what rules, steps and techniques apply specifically to you and your situation at every step along the way.

In the last chapter, we talked about having a realistic time frame. Don't let the contestants on *The Block* fool you. The teams are usually able to build the majority of a house themselves (with the help of their tradesmen) in just ten weeks, but that's not reality. If you're looking at doing it yourself, you might reasonably take six months or a year just for designing your house and deciding on all of the elements you'll need to consider before any work begins on your property.

Time spent on planning is worth it. If you have a watertight plan going into the project, you'll be more likely to stick to your schedule and your budget once your tradesmen arrive and the work begins. We'll talk about how to make a schedule and a feasible budget in Chapter Seven.

WHAT DO I WANT?

Before you can start thinking about what you want to build, it's important to first consider the broadest factors.

What type of home would best suit your lifestyle?

To work out what you need in a home, think about the following questions.

- How much time do you spend at home?
- Do you travel a lot for business or pleasure?
- Is your job or family likely to take you to another city or region in the near (or maybe not so near) future?
- Would it bother you if you put your heart and soul into building the house of your dreams only to sell up and move on within a few years?

If your home is only a place to make pit stops for you, you'll probably build differently than you would if you're the type of person who spends a lot of your time working, relaxing or entertaining at home.

What type of home would best suit the location you prefer?

Think about where you want to live.

- Would you like to live near the city, in the suburbs or out in the bush?

- Is proximity to the beach or the forest important to you?
- How important is view and outdoor space to you?
- How far are you from schools, shops and other services?

Where you choose to live will definitely help to determine what type of home you can build, both in terms of size, structure and materials.

What type of home would best suit your family?

Think about your lifestyle and what type of home you need.

- How many kids do you have or plan to have?
- Do you anticipate entertaining a lot of friends or family in your home?
- Would it help if you had enough room to accommodate grandchildren, either now or in the future?
- Would it be better for you to build a smaller home that's cheaper to heat/cool and easier to maintain?
- Could you really use a home office, a hobby room or a separate home theatre?
- Do you think you should prepare a granny flat for future use?

Though you can never know for sure, it's a good idea to consider how your family is likely to grow - or shrink - as the years go by.

How do you intend to use your property?

How you use your home should be considered in the planning stage.

- Are you planning to live in the house you want to build, or would you rather build an investment property that you can rent out?

- Do you think you'll live in the house for many years or do you plan to wait for the property to increase in value and then sell it and move on?

If you're building the house you think you'll live in for the rest of your life, you'll probably build differently than you would if you think you'll pull up stumps and move on in a few years.

What can I do with my property?

Keep in mind if you want to use your house as an investment property, there are some regulations regarding how long you have to live in your property before you can do anything else with it. Find out what the current rules are for your state.

What other factors may impact your decision to build?

Any number of other factors can influence your situation, and only you can know what those might be. The more factors you consider prior to starting, though, the more successful your project is likely to be, both in the short term and in the long term.

A Word to the Wise

Design for your life

The most important thing is to design your house around your needs. For my house, I knew I was going to have young kids so I wanted to be able to keep my eye on the children all the time, whether they were outside or inside. Also, I spend a lot of time entertaining friends and family at home, so I wanted to be able to spend time with my guests and manage cooking and such at the same time. I was able to design a house that allows me to do everything I want. With a lot of thinking and planning, you should be able to do the same.

WHAT TYPE OF BUILD WILL BEST ACCOMPLISH WHAT I WANT?

Once you have a good idea of what you want, now you need to decide the best way to go about building to meet your needs.

Renovate

One option is to expand or change the house you already have. You may love the location or there may be something special about your house that you want to preserve – it may be an old historical home, it may hold some special sentiment value or it may include unique features you can't replicate elsewhere. In this case, a renovation may be a good way to update your home and/or give you more space.

Also, you may be more or less happy with your home but you might just want to update your house to include a few energy-efficient features that probably wouldn't have been included in a property built ten or fifteen years ago. Adding these features is definitely worth the expense because, by doing so, you'll save money on utility bills and add value to your house.

The downside of renovating is that you will have less freedom in designing your space because you'll have to work around the proportions of the section of the house you want to keep.

On the other hand, even an extensive renovation can be less expensive than starting fresh (though sometimes renovations can cost even more than a new house, depending on what you want).

Build up or out?

A surprisingly easy way to expand your living space without cutting into your garden or yard space is to go up – by adding a second level. As long as the corners of the structure are stable and your foundation is strong enough, you may be able to design an extension that allows you to add another storey to your house.

First, you'll need to add posts through your existing walls to create a platform and remove all or part of your roof. Once you've created the platform, it's a simple matter of putting up walls – basically building boxes above the posts – and utilising that area how you want. The posts will transfer the load of the boxes through the internal walls and down to the ground.

You may have to pull down a few walls, but the process isn't as complicated as most people think. When you're ready, you'll have the option of rebuilding with a different style of roof or replacing a tile roof with a steel one or vice versa. Again, all those decisions depend upon your preferences.

The only obstacle to consider when building up is the envelope of your house. The envelope is the entire zone inside of which your home must fit, both in terms of perimeter and in terms of height. Your local council will determine how high you can build your house so that your structure doesn't obstruct the view of your neighbours. We'll talk more about the building envelope in Chapter Five.

What about THIS house?

So you've found a house you'd like to renovate. Before you buy, you should definitely get someone to come and inspect the property to make sure it's structurally sound before you buy. A carpenter or builder can look at the property for you, or you can get an independent inspection from a building inspector.

Keep in mind that an inspection (even a detailed one done by an expert) can reveal only so much. Before I built my own house, I bought an older property and initially, it was my plan to renovate. When I started pulling the house apart, though, I discovered that the house had termites and that a lot of the structure was rotten. I also realised that, with the existing roofline, I couldn't really design what I wanted. I tried a few options but in the end, I decided to bowl it over and start again.

The bottom line is that you can get an inspection done before you buy a house, but those inspectors can't always see everything that's going on with a property. They can't pull the plaster off the walls or pull up the floorboards. In other words, you're never really going to know what's what with a house until you start pulling it apart so stay flexible. Like me, you may decide after the purchase to demolish the structure and start over.

Buy a block and build

An easy option is to find a new block of land that no one has built on before and begin fresh. This is probably the easiest because there's nothing in your way, no services that you'll need to remove or turn off and no foundation to demolish. Also, the companies that provide utilities such as gas, water and electricity are always looking to attract new clients, so you may be able to get them to connect those services to a new plot of land free of charge. There are a few drawbacks to building on a fresh plot of land. Newly-developed estates sometimes lack the character of older, more established neighbourhoods, and they may be situated further away from town than what you'd prefer. Also, the price of a new plot of land may increase your overall cost to a point that your project becomes unfeasibly expensive.

Demolish and build

Want a brand new home but in an older, more established neighbourhood? It can be possible. Another option to consider is to find an old, reasonably cheap house in a more attractive location, demolish it and then build exactly the house you want on the plot of land.

It may not seem cost effective to buy an old house only to bowl it over and rebuild, but doing so may increase the overall value of the property exponentially. In general, the cost for demolishing a house can be quite low - generally in the vicinity of $10,000 or $15,000 - and it usually only takes one or two days. Do the math and see if demolition is a good choice for you.

TYPES OF FOUNDATION

One important element to consider is the type of foundation that will best work with the design you have in mind. If you're renovating, you may be able to build on top of your existing foundation or simply reinforce what's already in place. If you're building new, however, your choices will probably be either a **slab**, **stumps** or **brick piers** - and there are pros and cons for each.

A **slab** is made from a single, thick layer of concrete that gets poured all at once to create a solid base on top of which to build your house.

Stumps are made from a series of posts (usually made from concrete or hardwood) that have been embedded into the ground and set in place with concrete.

Brick piers are made from a series of pillars made from bricks that rest on top of a concrete pad.

Slab

These days, a lot of architects will automatically design a home to include a slab. It's quick and easy to install and comes with a few benefits.

One benefit of a slab is that there is absolutely no movement, and many people find this level of stability reassuring. In some places, particularly where termites are prevalent, the law requires the use of a slab instead of stumps.

Secondly, in certain climates, a concrete slab can soak up the heat during the day and help keep your house warmer at night. How well a material can absorb and hold heat is called **thermal mass**, and a slab generally has a lot of it. If you live in a region where the temperature tends to change significantly from day to night, the thermal mass a slab provides can work to your advantage.

Thermal mass refers to how well a material can absorb and hold heat – a slab generally has a lot of it while stumps and piers have less.

On the other hand, the cost of a slab is nearly double that of the alternatives. Using a slab will also reduce the flexibility in your design. Once that slab is poured, your pipes will all be frozen in place and you will be quite limited as to what you can change.

Another consideration is possible repairs in the future. Most slabs are generally very solid and will withstand a lot of wear and tear, but if something ever does happen to your slab, the cost of repairs can be significant. You'll also have no access to whatever utilities are lying under your house, and that lack of access can be a problem if other maintenance issues arise.

Stumps

One alternative to a slab is *stumps*. With stumps, you'll begin by digging several holes, usually fifty or sixty for a whole house at a depth of 900 mm, although these numbers will depend entirely on what you're building. These holes will then be filled with concrete (usually) to create a series of solid posts on top of which you will build a **subfloor**.

The **subfloor** is the layer of flooring that is built over a stump or brick pier foundation, creating a flat, solid surface.

The benefit of using stumps is that they make it possible for you to move things later on and they cost generally around 60 per cent of what a slab will cost. With stumps, you will get a bit of movement, but sometimes you want some flexibility.

Stumps are usually the best option (sometimes the only option) when you have a plot of land that slopes, for example. And if you happen to experience an earthquake – which happens more often than you might think, especially in Victoria and New South Wales – even a mild one can cause serious damage to an unyielding slab.

Once a slab is poured, it is very difficult to make any structural changes in your design. Stumps create a cavity between your house and the ground beneath it. They allow you to move walls and pipes, either at the time of the build or years later, depending on what you or future occupants want.

For obvious reasons, stumps will have less thermal mass than a slab. It's an easy task, though, to insulate your subfloor so that your stumps can be as energy efficient as a slab but with more flexibility and at a lower cost. Also, while stumps may require more regular maintenance than a slab, the cost of replacing or adding stumps is substantially less than the cost of repairing a slab.

Whether or not you can use stumps will depend in part on the type of soil you have and where you live. Your engineer can tell you whether or not stumps are possible and your architect will discuss with you the feasibility of using stumps with the design you have in mind. It's a personal thing, but personally, I prefer stumps. We'll talk more about the architect and engineer in Chapter Five.

Brick piers

Another foundation that allows for flexibility is brick piers. Very similar to stumps, brick piers also create a cavity beneath your house, which allows you to move pipes and other services when necessary. In some states, where they don't use stumps at all, brick piers are the alternative to a slab.

With brick piers, you'll start with a concrete pad and then build piers - or pillars - out of bricks and mortar on top of them. You'll then install bearers and joists to create a subfloor the same way you would with stumps.

Brick piers have many of the same benefits and drawbacks as stumps. They do take longer to put in than stumps because they must be built one by one, and they can be more difficult to protect from termites. Brick piers also tend to be a lot more expensive than stumps and, with so many components, there's a lot more room for error.

A Word to the Wise

See what you can dig up

When preparing the ground for your foundation, be prepared to find anything, just like I did when I was doing my garage foundations. We had dug all of the beams and footings but as we were getting to the last stage of the foundation, we noticed a big wheel - the type you'd see on a cart meant to be drawn by a horse.

We kept digging and we ended up finding the suspension and the frame for the cart. For me to pull the cart out would have meant destroying all of the **excavation** we'd done for the previous two days, so rather than remove the entire cart, we just cut away the bits that were in the way. I figured the cart had probably already been there for a hundred years - no harm in leaving it there for another hundred.

Excavation means digging out the foundation for your house, usually with a large digging machine like a Bobcat.

Saving your existing foundation

If you're planning on renovating a house, there's a chance you might be able to build on top of the existing slab or subfloor, but I wouldn't recommend it.

For one, keeping the existing foundation will limit your design. You've always got to have adequate structure underneath your house to cope with the weight of the walls and the roof. If your stumps or slab are not positioned properly, the integrity of the entire structure will be in jeopardy.

Also, an old foundation is more likely to deteriorate sooner than you might hope. The last thing you want is to budget for a build only to discover your foundation isn't up to scratch.

It would be the rare occasion when saving the foundation would be the best choice. For the most part, I would recommend pulling out the old foundation and replacing it with a new slab or stumps. The exception to this rule is when you're renovating and you plan to keep the majority of your house. In this case, you'll still want to make sure the foundation you have in place is adequate to the task you have planned.

If you do decide to keep the existing foundation, you'll need to be sure and demolish the remainder of the house properly. We'll talk about methods of demolition in Chapter Eight.

TYPES OF FRAMING MATERIALS

Now is also a good time to consider what type of building materials you might want to use for **framing** your house,

although it will be worth talking with your carpenter/ foreman, architect and engineer first, and considering your budget before making any final decisions about materials. Framing generally refers to the walls of the structure, or sometimes the walls and the roof.

The **frame** of your house is the interior structure of your walls and sometimes also your roof. Usually the frame is covered with some other material.

Building these days is better than ever. There's more science in building and with each step along the way, you will have a lot of different choices, including materials that can create lightweight, well-insulated structures that may be cheaper than traditional building materials.

Solid brick

Brick is undoubtedly sturdy and generally well insulated, but it's not necessarily your best option for a lot of reasons. For one, it takes brick a lot longer to heat up but also a lot longer to cool down - which is great in a cool climate but not the best material in a warm climate or during the summer.

Also, brick is very heavy which means it needs a stronger (and usually more expensive) foundation to hold it up. And it's expensive - often significantly more expensive than most of the alternatives - and it takes a long time to install. For all its drawbacks, though, brick is still the strongest of the building materials and stands up really well against weather, termites and other pests.

Brick plus rendering

If you really like the look of brick but would like to save some money, one option to consider is brick plus **rendering**. In contrast to solid brick - in which your house is basically walled by two layers of brick - this method combines one layer of brick with a second wall built from a different material.

Rendering means covering another type of material with plaster to create a smooth surface.

With this technique, you add a layer on the inside of the brick, creating a skin of insulation that is equally effective - if not more so - than solid brick. The downside is that whatever material you use for the non-brick layer is likely to be susceptible to termites, but it will probably be a lot less expensive than two layers of brick.

Do I really need to worry about termites?

A lot of today's modern building materials are designed to stand up to termites - you can even buy timber that's been treated so that termites just won't touch it. Many areas don't really have termites but it doesn't hurt to include measures to keep them out, even if your area isn't considered to be a termite risk.

Cement sheet

Some people really like using cement sheet, but personally, I wouldn't recommend it. When you install lightweight cement sheet to the frame of your house, you'll probably save money - but because it's so flimsy, you're likely to start seeing cracks within a year or two and you'll almost certainly have constant maintenance issues.

Concrete blocks

Another one of those materials that might not come automatically to mind is concrete blocks, but they can be an excellent option. For one, they're fantastically strong and if you render over the top of them or paint, you'll get the same look for a much lower cost.

Basically, concrete blocks give you the strength of brick but they're cheaper and quicker to install. The only thing you lose is the appearance of the bricks, but keep in mind that, if you're going to go with solid brick or even brick veneer, you'd better love it because once those bricks are laid, you're stuck with that look. However, with rendered concrete blocks, it's a much smaller task to repaint your house a different colour every few years if you want.

Polystyrene

It might not seem like the best material to use when building a house, but it's one I would really recommend. For one thing, it takes very little time to install. More importantly, a single layer of polystyrene 75mm to 100mm thick will give your house a thermal rating of about 3.6, which is the minimum you're trying to achieve. Add insulation and that thermal rating goes up to six.

In the same way that polystyrene cups keep liquids hot or cold, it can do the same for your house. You can turn on your heater for an hour in the morning, get your house up to 21 degrees and then turn off the gas – the house will still be warm when you get home at five o'clock, even in the middle of winter.

One of the drawbacks of polystyrene is that it can be penetrated. A standard wall frame is usually built with 450mm between each **stud,** which means a person could break through a polystyrene wall using nothing more than a shovel. The same, however, is true of weatherboards and other building materials that don't provide the same benefits of polystyrene.

A **stud** is a piece of timber that is used to create a frame. The long studs stand upright and generally run from floor to ceiling.

You can solve the security concern of polystyrene by putting in extra studs so that there's only 200mm between the studs, which means if a person broke a hole through that foam, they still couldn't get into the house.

Brick is definitely sturdier than polystyrene, but it can cost twice as much and you may not end up with a superior product for the price. It just depends on your individual needs and what you feel comfortable with.

Steel

Another good material to use for framing is steel. For one reason, termites are no issue at all. Also, steel frames have been around for a long time, but they've now been engineered to eliminate the squeaking and other noise you used to see in older steel frames. Steel frames have all the benefits of timber frames but they're very strong, they're perfectly straight, they go up quickly and they can be cheaper than other options.

Give your plans to a company like Supaloc, and they can manufacture your house in their factory and deliver the completed frame to your property. The entire frame will probably go up in only a day or two, and you might be able to save yourself a week or two in the build.

A steel frame will also come with all the service holes pre-drilled, which saves time for your plumbers and electricians, who don't have to mess with drilling holes into the structure of your house - and you eliminate the risk of one of those tradesmen making a mistake and ruining your frame.

Alternative materials

You might be amazed to hear some of the cutting edge materials that people are using for building these days, but sometimes the most unlikely substance can make a fantastic structure.

Would you believe, for example, that houses made from bales of straw is a thing? It should come as no surprise - it's a method that's been around for centuries, and some straw-built houses have managed to remain standing for hundreds of years.

Another option is using mud. As with straw bales, mud-brick houses have been around for centuries. Both straw and mud provide excellent insulation so that the interior of your house stays a constant temperature, summer and winter. If you're particularly interested in sustainable building methods and materials, it's worth doing some research to see what options are likely to work best for your area.

When it comes to choosing materials, the choices are basically endless. It's simply a matter of choosing your design and then working out the cheapest way of building it. There are many cost variables but you have to weigh up what's important to you and where you want to spend extra money. If your budget is limited, you might be able to save money on materials. You may also want to consider if you'd rather pay a little more for materials that are easier to maintain.

TYPES OF ROOFING MATERIALS

Another important consideration is what type of roof you want. In general, most people use tiles, steel sheet or sometimes shingles.

Tiles

There are basically two types of roof tiles – terra cotta and concrete. There was once a time when people thought terra cotta was better, especially in the cold or wet parts of Australia, but with the improvements that have come along in recent years, the two types of tiles are about the same these days.

The benefit of using tiles is that they're very strong and durable. You can expect a roof that's been tiled properly to last at least 50 to 100 years. Another plus (for some people) is that roof tiles do a much better job in reducing noise.

You're also not very likely to experience many problems with tiles. Because each is laid individually, you can replace just a few if you need to rather than replacing the

entire roof or even large sections of it – which is especially useful if a large gum tree drops a branch on your roof and causes some damage. In general the maintenance is very easy. At most, you'll only need to have your tiles resprayed maybe once every ten years depending on how much sun and rain you see.

At the same time, tiles are strong because they're heavy, and a heavier roof means you'll need a stronger foundation and a stronger frame. And as is usually the case, the heavier the materials are, the more they cost. You'll also pay more in labour to have each of those individual tiles installed one by one.

Steel sheet

An alternative to tiles is the old Australian favourite, the steel sheet. Though metal roofing used to be made from flimsy corrugated tin, today's steel sheet roofing materials are excellent. A lot of people prefer the traditional look of a steel roof for the same reason they look for convicts in their genealogical records – just for the history of it. We've been using steel sheet roofing in Australia since the 1840s.

Cost is also definitely a factor in choosing a steel roof. Because it's lightweight, steel sheeting doesn't require the same level of strength in the foundation and frame, which means you can save money literally all around your house. You also have nearly endless options in terms of colour, it's hail and impact resistant and it's easy to maintain.

While a lot more durable than the tin roof sheeting of old, modern steel sheet still isn't as durable as tile. Most steel roofs come with a 25-year warranty and you can expect to see flaking and chipping after 12 years or so.

Keep in mind, too, that in a heavy rain, that steel sheet will make a lot of noise - which is a pro for some and a con for others.

Shingles

Though they're a lot more common in America, many Australians have started considering asphalt shingles for their roofing. Sort of a cross between tiles and steel, shingles are lightweight but durable. They usually come with a warranty good for twenty to forty years, they're easy to insulate and they're very versatile in terms of look.

If you think shingles might be right for you, talk to your engineer and your foreman. We'll talk more about the engineer in Chapter Five and the foreman in Chapter Six.

THE BUILDING PROCESS – A SNAPSHOT

At this stage, it's a good idea to take a look at the process you can expect to go through in building your house from beginning to end. These are the steps we're going to cover in detail as we go through the rest of the book.

Design your rough plans

This is the fun stage - when you get to dream up every single design aspect you'd love to include in your home, from top to bottom, inside and out. You have a lot of options to consider including rooms, features, techniques and materials. Just keep in mind that some of these ideas may not survive all the way to the end of your project.

Get your financing and your documents in order

Now you'll need to find a lender, get pre-approval so that you know how much you'll have to spend, obtain all the necessary permits and get all the different types of insurance you'll need to stay covered throughout the project. The rules are different in every state, but these guidelines will give you a starting point.

Organise your land surveyor, architect and engineer

Next you need to find out the exact boundary lines of your property from a land surveyor then have your rough plans drawn into proper **blueprints** by an architect. Then your engineer will work out how to make it all happen. This is the point at which you will start to finalise your design.

Blueprints are the precise technical drawings an architect will have a draftsman draw up – your tradesmen will refer to these drawings in building each step of your house.

Organise your carpenter/foreman

For the amateur owner builder with very little construction experience, the foreman is one of the most important people you'll hire throughout your entire build. Make sure you choose carefully because you will need to rely on this person from now to the very end of your project.

Organise your tradesmen and building inspector

With the help of your architect, your engineer and/or your foreman, you can now find the contractors you want to use and obtain quotes for labour and materials. With all of your quotes in hand, you can draw up a feasible master schedule and budget, and then finalise your financing.

You'll also need to organise the building inspector, the person who checks out each phase of your project.

The foundation stage

You're almost ready to start building but first, you need to prepare your job site so that it's safe and comfortable for everyone. If you're renovating, you may need to demolish an old structure. Then you can oversee the excavation, pouring and curing of your foundation, which will probably be a slab, stumps or brick piers. This stage will include an inspection after excavation.

The framing and lock-up stages

Now you're ready to put in a subfloor (if you're using stumps). Then you can put up the basic structure of your walls, put on your roof and then **lock up** by installing your windows and external doors before making sure all the cables and pipes are in place for your utilities. This stage will also include inspections of the frame followed by **cladding**.

The **lock-up stage** is when the subfloor, all the exterior walls, the roof, the exterior windows, doors and cladding are put in place. In other words, the build has progressed to the stage where you can lock out prowlers and the weather.

Cladding is the façade of your building, the external layer that will be visible on the outside of the house and covers the internal frame of the structure.

The fit-off stage and the finishing touches

Now it's time to install all of the final fittings, plastering, painting and landscaping. At the end, your inspector will return for one final check. Pass this inspection and you'll get your Certificate of Occupancy which allows you to move in.

You'll also want to consider the finishing touches such as how you want to decorate your home and what to do to maintain it so that it continues to look as good and work as well as when it was first built.

How much building will I get to do?

As an owner builder, you'll want to keep an eye on everything that happens on your project because the buck stops with you. Once the construction starts, there will be plenty for you to do, even if you do follow my advice and hire qualified tradesmen to do most of the actual building. A lot of these jobs won't be very glamorous or technical, but they still need to be done and if you don't do these jobs yourself, you'll have to pay someone else to do them. We'll talk more about what you can do to help in Chapters Eight, Nine and Ten.

CHAPTER SUMMARY

- The planning stage could - and probably should - take more time than the rest of your build combined.

- In deciding what you want to build, consider what type of home is best suited to your lifestyle, what your ideal location is and what your needs are including any family needs. How you intend to use your property and any other factors will also influence what type of house you build.

- Keep in mind that, as an owner builder, you'll probably be required to live in your house for at least a few years before you sell. Find out the rules for your state.

- You also need to decide if you want to renovate an established home, build on a new block of land or demolish an existing structure and rebuild.

- If you really want more space but you don't want to cut into your yard, consider going up by adding a second storey to your existing house.

- Be sure to get any property thoroughly inspected before you buy.

- Now is a good time to think about the type of foundation you might want to use - a slab, stumps or brick pier. There are many pros and cons to each, but I personally prefer stumps.

- If you're renovating, you may be able to save the existing foundation and/or subfloor, but it's not usually the best option.

You can also start thinking about possible framing and/or cladding materials you may want to use, including (but not limited to) solid brick, brick plus rendering, cement sheet, concrete blocks plus rendering, polystyrene with rendering, and steel. There are also other alternative materials such as straw and mud that are particularly sustainable.

While you're at it, think about your roof. Your three main options for materials are tiles, steel sheet and shingles.

At a glance, your project will include a few steps, starting with drawing up your rough designs and organising your finance and documents. Then you'll hire a land surveyor, architect and engineer plus a carpenter/foreman, all of your tradesmen and a building inspector. Finally it's time for the foundation stage, the framing stage, the lock-up stage and the fit-off and finishing stage. Then we'll look at what to do after you move into your new home.

As an owner builder, you will always be able to do some jobs around the site once construction begins.

WHAT COMES NEXT

You've got a general idea of what you want. Now it's time to put pencil to paper and get more specific about the design of your house – just keep in mind that you'll need to stay flexible until you've sat down with an architect and that you may continue making changes right up to the very end.

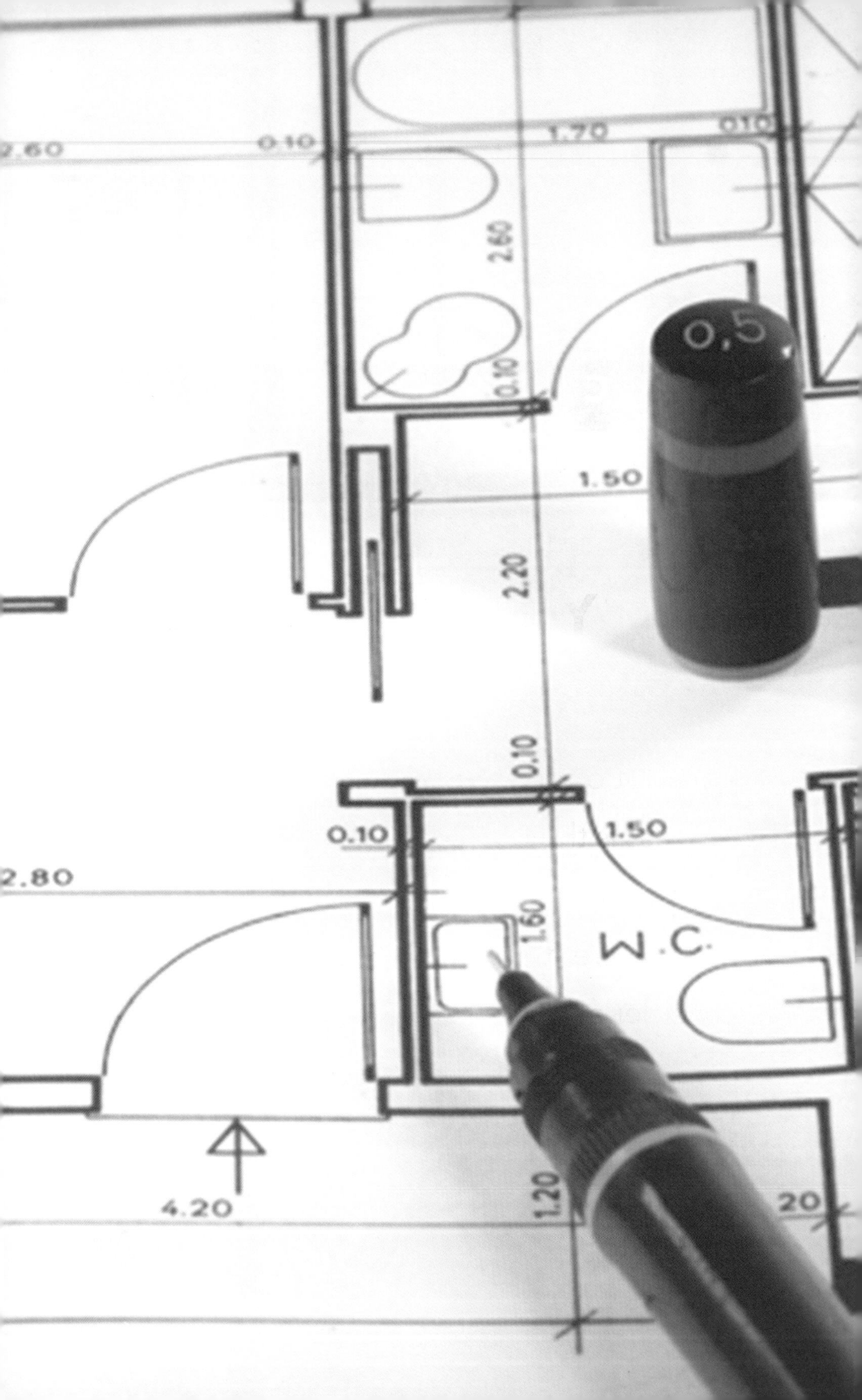
0.10
1.70
2.60
0.10
0,5
1.50
2.20
0.10
0.10
1.50
2.80
1.60
W.C.
4.20
1.20

CHAPTER THREE

DESIGNING YOUR HOME

For a lot of people, this part of the planning process is the most fun. Now is when you get to choose every element of the home you want to create. There are a lot of options to consider at this stage – some of which might not have occurred to you before.

HOW WILL MY DESIGN BEST WORK WITH MY LAND?

Start by looking at your land – both the size and orientation of your house will depend upon the size and lay of your plot of land.

How big is the entire plot? Make sure you're completely certain about the boundary lines between your property and those around you, and find out how much space you will need to allow between the walls of your house and the boundary line. To find out, you'll need to get the dimensions of your property measured exactly, including how high it can go. We'll talk more about getting your land surveyed in Chapter Five.

Once you know how big the perimeter of your house can be, consider the best way to position your house to make the most of the landscape. Where does the sun come up and set? North facing is ideal, but don't worry if your land isn't north facing. You can always position windows and include skylights to allow natural light into your house at the appropriate times of day and year.

Does your land slope significantly? Is there a particular view you'd really like to maximise? Think about how the position of the house will make the best use of the features your land has to offer. How will you maximise your yard space? If you have children, you may want to keep more yard space by building a second storey.

If you'd rather cut down on the amount of gardening you'll need to do, you may want to build closer to your boundaries and reduce that yard space. Either way, you should think about it before you start.

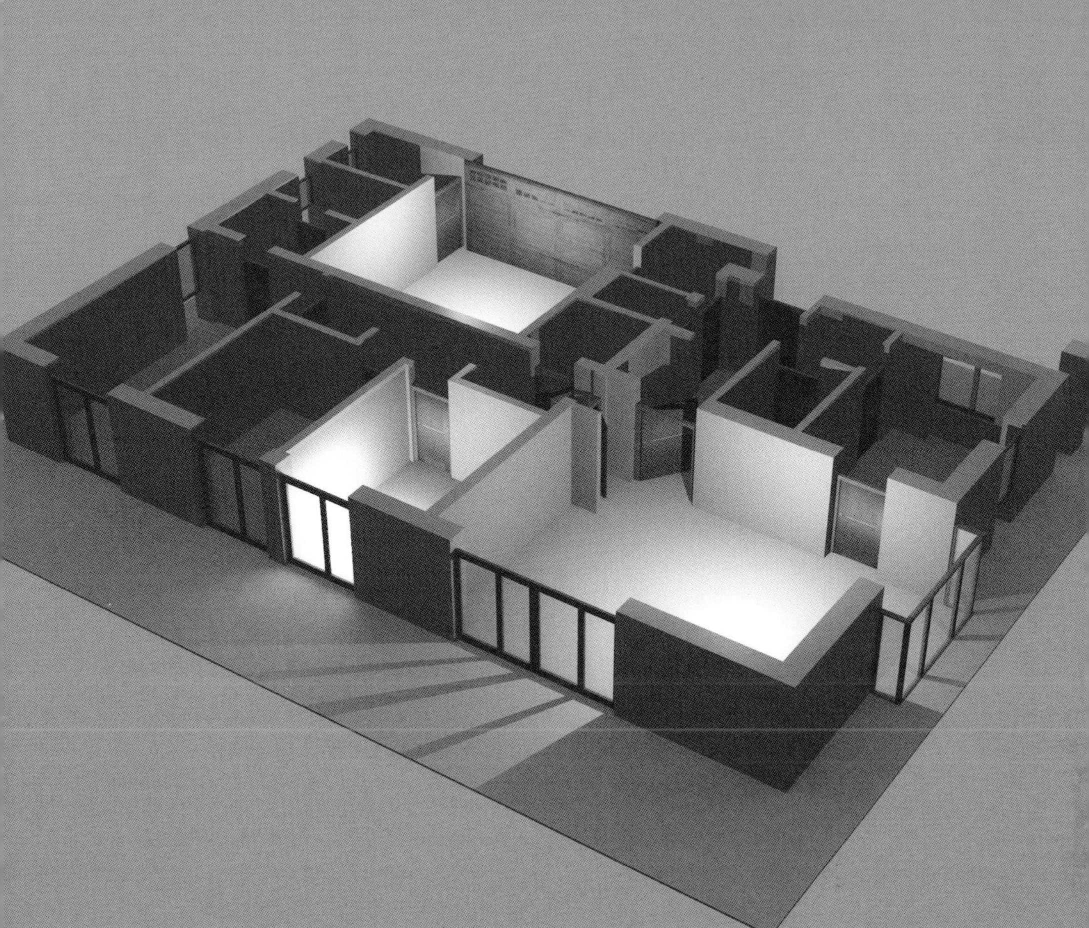

EXACTLY WHAT WILL MY HOUSE LOOK LIKE?

When it comes to designing your ideal home, the Internet is obviously an invaluable asset, but it's not the only one to consider. There are dozens of design magazines available at your local bookstore or grocery store, all chock full of amazing ideas. There are even a few TV shows all about renovating and building. (I can recommend a good one!)

More than just looking at professional photos of other people's ideas, the best place to start in designing your ideal house is by checking out your own life.

Look at where you're currently living. What do you love about it? What are you lacking? Do you like a lot of light? Do you like more privacy? Do you like a lot of space? Do you like high ceilings? Do you want to include outdoor spaces for entertaining or an enclosed sunroom? Do you expect to spend a lot of time in the kitchen? Think about how you can best incorporate these elements in your design.

Think about how you'd most like to use your space. Would you prefer your children share a room or have their own rooms? Do you plan to store toys in the bedrooms, or keep the sleeping spaces smaller and include a second lounge room or play room for toys? Do you want one big bathroom or a few smaller bathrooms? Would you rather have one big open plan or divide up your lounge space into separate areas?

How important is storage to you? If you're the type who hates clutter, now is the time to incorporate plenty of closets and cupboards where you can neatly stash away everything in its own proper place. Consider how you use your current space – where would be the perfect place for an extra cabinet or set of shelves?

Looking for ideas in terms of façade and landscaping? Go for a drive. There's nothing wrong with looking at houses in your own neighbourhood or others where you like the look of the homes. Even better, you can mix and match all of your favourite elements from as many different homes as you like.

Not sure what you want? Think about rooms you've seen or been in before and loved in the past and consider how to integrate those ideal elements into your own design. Visit display homes by volume builders, look at pictures in magazines and online. Drive around and take photos of façades that appeal to you. There are literally thousands of options to consider – it's just a matter of choosing what you like for the amount of space you have.

Q&A

What are some pros and cons of high ceilings?

A lot of people like high ceilings, including me. I personally love high ceilings because they tend to give just about any room a feeling of grandness. They also allow you to include more windows high up on your walls, letting in more of that natural light that creates atmosphere and saves on lighting bills. A high ceiling can also make a small room seem bigger.

At the same time, one of the most basic laws of thermodynamics is that heat rises. If you live in a warm climate, you may want the heat in your rooms to rise. If you live somewhere that tends to be colder for most of the year, you'll want to keep that heat closer to the floor. The good news is, if you live in a colder climate, you can help push down the warm air in a room with a high ceiling using ceiling fans.

You should also keep in mind that building a high ceiling may cost a bit more, and if you need to do any maintenance on fans or light bulbs way up in that ceiling, the process will be more difficult than it would be in a standard height ceiling. Getting rid of cobwebs high up can also be difficult.

Despite the drawbacks, I'm still a big fan of a high ceiling - but as always, it will be down to your personal preference.

ROOMS YOU MAY WANT TO INCLUDE

What rooms you want to include in your house will depend on what you like, what you use and your budget. A good idea is to think about which room in your house you use the most – where you spend the majority of your time or where you spend your favourite time – and build around that room. It doesn't hurt to consider all of your options before getting started.

Living room

For most people, the living room – or lounge room – is where they spend most of their waking hours when they're at home. The living area might be formal or casual depending on your preference. It might be a large open area or a smaller, more discrete room. Create this space so that you love it and it accommodates the way you relax.

Kitchen

For a lot of people the kitchen is the hub of their house. No matter what is happening, it seems like the kitchen is always full. Think about how you use your kitchen. If you do a lot of entertaining, you may want to create a space that allows you to cook or prepare while spending time with your guests at the same time – like I did. If you hate to cook and you spend more time going out than entertaining at home, you might prefer to put your resources elsewhere.

Dining room

As with everything else in your house, how much you use a separate dining room or meals area will depend on your lifestyle. If you like to keep the mess of preparation out of sight during your meals, a separate dining area from the kitchen would be practical. Or you might prefer to cook while keeping an eye on what's happening around the table at the same time.

Office/study

If you work from home, you might really benefit from having a separate office, or you might prefer a room devoted to crafts or hobbies. There are tons of ideas for creating an artist's studio or writer's retreat within your home or even as a separate freestanding building elsewhere on your property.

Media room

How good would it be to have a separate room designed just for watching movies, sports and other favourite TV shows?

A lot of modern homes now include a formal lounge room for entertaining and a different room just for the TV or a second television.

Games/play/rumpus room

If your kids are like mine, they probably make a huge mess with their toys pretty much every day. Sometimes they want to return to a particular game the next day.

Rather than having toys strewn throughout your house all the time, consider designating one room or space exclusively for toys and keeping them all in one place. Then you can shut the door and the rest of the house stays tidy.

Bedrooms

If you're the type who likes to read or watch TV in bed, you'll want to make sure your bedroom is a comfortable space.

Consider everything that goes into creating a calm sleeping environment. What do you have right outside your bedroom window? Do you want privacy or do you need to be close to children? How well does your en suite and closets suit your daily needs?

You may need to set up your space so that you don't disturb your partner when you get ready for the day in the morning or when you come home at night.

Think about your lifestyle to design your bedrooms in a way that will suit your household needs.

Bathrooms

When it comes time to sell your house, your bathroom can make or break the sale and for good reason. A well-appointed bathroom with plenty of storage and bench space can make all the difference when it comes to getting ready or for bathing the kids comfortably and efficiently. You may want to think about housing the toilet in a separate room from the shower or installing a spa bath if you like to linger.

Hallways

There are lots of interesting ways to use hallways. Think about incorporating floor to ceiling bookcases or other storage so that you can make the most of that space.

Garage

You can use a garage for many things. You might park two cars or park one and use the rest of the space for storage, so think about what you need from your garage.

How do you want to access your garage – from your front door or from an interior side door or both? Do you want a separate garage or one that's attached to the house? Would you use it more as a workshop?

You might find people are constantly coming and going from your house – you don't want to design your space in such a way that you're constantly parking in someone else and having to move your car.

Granny flat

Think you may need to look after an elderly parent sometime in the near future? You may want to include a **granny flat**, which is usually a freestanding apartment or studio built on your property, perfect for an aging person who wants their privacy but may need some looking after. You might also want a granny flat where a university student in your family can live.

A **granny flat** is usually a freestanding apartment or studio built on your property. It can come with more or fewer amenities, such as bathrooms and kitchenettes, depending on your needs.

Granny flats can come with more or fewer amenities - such as bathrooms and kitchenettes - depending on your needs.

Other speciality structures

What other special areas do you want to create: a sunroom, deck or pergola, cellar, workshop, swimming pool or tennis court?

You know what you love and how you spend your time, and your house should be designed to cater to your lifestyle. Think about what else you want to include outside so that you can incorporate those designs as efficiently and as cost effectively as you can. More about these structures later.

A Word to the Wise

Get a feel for the space

Sometimes it's hard to know what the size of a room will feel like based on a picture or a floor plan. The last thing you want is to build your house and realise that the space you planned is actually a lot smaller than what you envisioned.

If you can't visualise the spaces, find a room with the same dimensions of what you have in mind so that you can see and feel what a room that size is actually like in real life.

DESIGNING YOUR PROPERTY INSIDE AND OUT

In designing your house, you need to take into consideration what's happening both inside and outside your house. If you don't design everything in the beginning, you may find that adding those outside elements after the fact will undermine the original structure of your house.

Think about all of the decks, pergolas, fences, workshops, swimming pools and other outdoor structures you want to include. These structures will all require foundations of their own and some will require electricity, plumbing and sometimes gas – and all of these services will need to be in place for your exterior structures long before the majority of the house is finished.

Your electrician, for example, may need to run all the power cables and leads through your house to get them in position for your outdoor lighting. You might also want external heaters, remote control gates or water features that might require power. In other words, get these elements into your plans as early as you can.

A Word to the Wise

Plan your outdoor structures

I once saw a builder who didn't plan for a cellar. It was designed to be about 400mm away from the wall of the house, but the builder didn't start thinking about how he was going to incorporate the cellar until after the walls of the house were up. He realised too late that putting in the

cellar after the fact was going to cause the house walls to fall over. It took a lot of work and many thousands of dollars to create enough foundation support to put in that cellar. If he'd built the cellar from the start, it would have been a much cheaper, easier task.

Your architect will be able to tell you what's feasible and what isn't before you start building. You may have your heart set on certain design features that simply aren't possible for the space you have.

Be prepared to listen to the advice of the experts – they'll help you find a solution to whatever issues may arise.

A Word to the Wise

Remain flexible

It's best to have a solid plan in place going into the project, but you may not be able to include every room, material or structure you'd like.

Your entire project will depend upon your financing. You may have a design in mind that simply isn't doable with the amount of money you have available. Having said that, there are sometimes options in terms of building materials that may allow you to build what you want with the money you have. You'll be able to draw up a more specific budget after you've spoken with your lender (Chapter Four) and you've organised your tradesmen (Chapter Seven).

DRAWING UP ROUGH FLOOR PLANS

Before you hire an architect to draw up blueprints and the rest of your plans, it's a good idea to draw up your own plans so that you can see on paper what you have in your head. Your rough sketches will also give the professionals something to go on when the time comes. We'll cover working with the land surveyor, architect and engineer in Chapter Five.

What you'll need to draw up your ideas

The following materials will help you to provide your architect with a clear picture of the design of house you want.

Sketch paper

Probably the very easiest place to begin is with a simple sketch. You don't have to be a great artist to plot out a rough draft of which rooms you'd like to go where and how the house will look on your land.

Got no idea at all? Take a look at floor plans on offer from the big volume builders – available online, in brochures or even in the local newspaper – to get a general idea of what the final draft will look like (but make sure you don't copy their plans exactly).

Graph paper

If you're still in the sketching stages, you can use graph paper to measure out your entire property, including land and house, so that each square of the graph represents one square metre (or more or less depending on your needs). Now design each room to scale.

Keep in mind the slope of your land, any views you want to capture and the direction of the sun. These plans won't be exact, but they'll help you visualise what you have in mind.

Online

If you really want to, you can purchase floor-plan software, but there are heaps of free websites that will help you design what you want. Search for 'make your own floor plans' or 'design your own house' and you'll find dozens of options. Pick the one that works best for you and experiment until you're able to develop your dream design.

Q&A

Can I copy someone else's plans?

It's a good idea to look at as many different options as you can when designing your house. Keep in mind, though, that it's illegal for you to copy the professional plans of another builder exactly. There are copyright laws for intellectual property that prevent you from nicking someone else's plans, especially those of volume builders. You can get ideas, of course, but make sure your plans are your own.

BUILDING AN ENERGY EFFICIENT HOME

Though it may cost a little more in the short term, including energy efficient elements into your initial design can save you thousands of dollars in the long run – and you'll probably save money incorporating these components into the original build rather than waiting to add those elements after the house is built.

- **Windows:** Double-glazed windows are expensive, but they make a huge difference in terms of keeping out the noise and keeping in the heating or cooling you want. You can even get windows with obscuring glass or tinting that lets in the light but reflects the sunlight and is able to keep out prying eyes.

- **Insulation:** Having your house really well insulated will save you a lot on heating and cooling costs. There are lots of options in terms of building materials that can keep your house toasty in winter and cool in summer. We'll talk more about some of those choices later in this chapter.

- **Seals:** Make sure the house is properly sealed so that there are no gaps around any of your windows or doors. There's nothing worse than gaps in your structure that let in a draft. An airtight home will reduce changes in temperature and save you money on heating and cooling.

- **Natural light:** Your local council will almost certainly want to see that you've included a lot of natural light in your design. Skylights and north-facing windows will save you money on lighting and allow in a bit of heat. Natural light also provides a lovely atmosphere.

- **Water tanks:** It may seem obvious, but rain is a fantastic free source of water, especially in areas prone to drought. You can utilise your roof by including gutters that drain into water tanks. Using that water for gardening and washing cars will benefit the environment and create substantial savings for you.

- **Solar panels:** How would you like your power company to send you a cheque instead of a bill? Solar panels are even better today than they were just a few years ago. The engineering has improved and the price has come down. They need to be north facing and, ideally, you don't want to see them from the street, but it's definitely worth it to include as many solar panels as you can in your initial design.

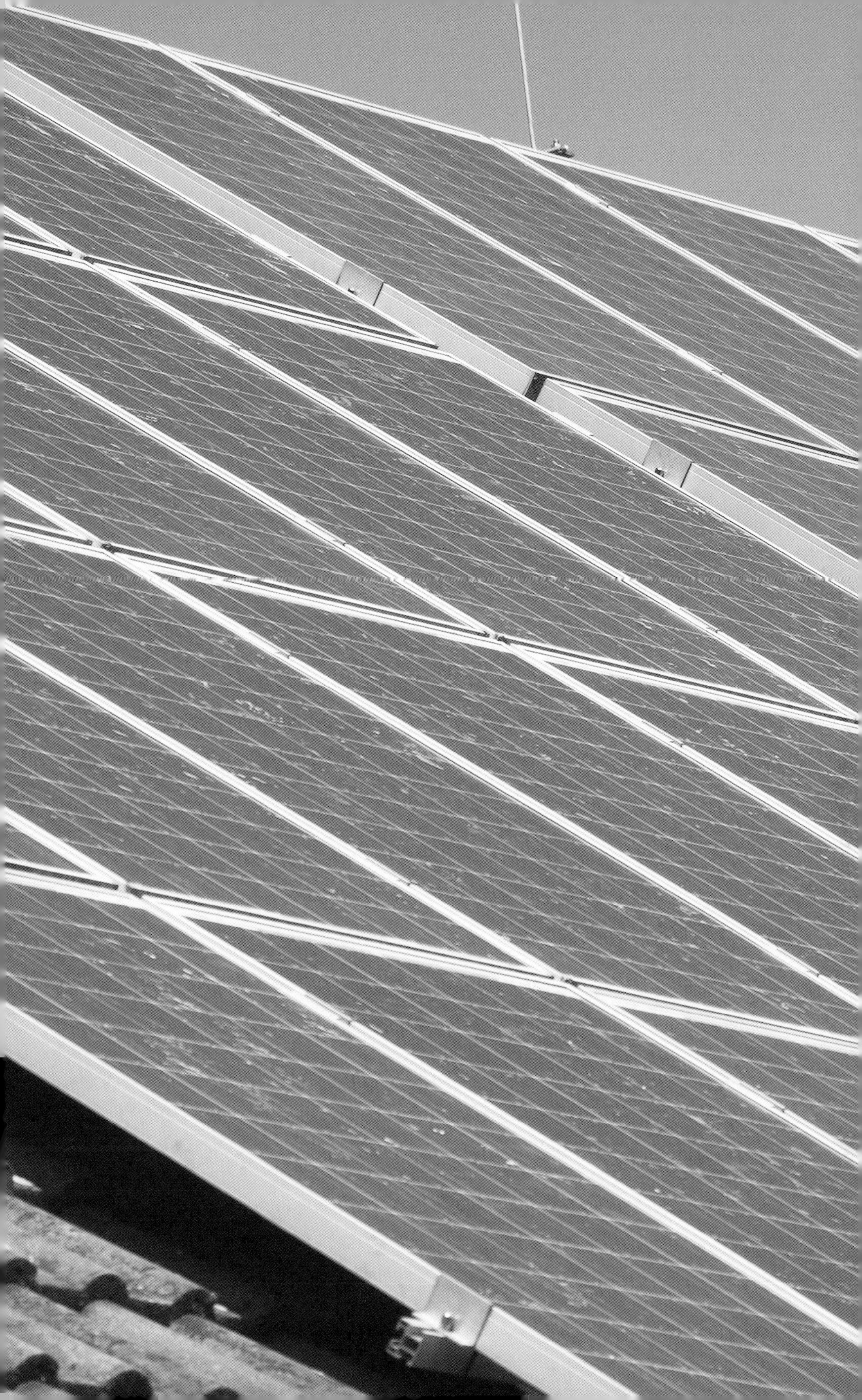

What constitutes a six-star energy rating?

The laws regarding the six-star energy rating have been a bit up in the air recently. Some states have been quite specific about what they require in terms of which energy-efficient elements need to be included in each newly-built home and to what standard.

If you're not sure, don't worry – your architect and engineer will make sure your home is designed to adhere to all the most current rules, and your inspector will insure that all the standards have been met before allowing you to move into your new house. We'll talk more about the roles of the architect and engineer in Chapter Five and the building inspector in Chapter Seven.

Can I add energy-efficient elements after the fact?

You can add things like solar panels and water tanks to established homes, but it's easier to do it when you're in the process of building the home. We'll talk more about why in Chapter Nine.

It may be tempting to save money on the initial build and to consider putting in energy-efficient elements later, but it will actually cost less to do it during the build.

If you're renovating, however, it's worth the money to make your house as energy efficient as possible, even if it wasn't built that way to begin with – you'll add value to your house and save stacks on your utility bills in future.

What time of year should I build?

Another important element to consider is the time of year when you want to build. Once your house is framed and locked up, the tradesmen can work inside without much concern for the weather, but rain can cause serious problems during the foundation and framing stages. In general, dry weather is best for every aspect of the build until you get to the very end. The ideal time of year will depend on where you live, but aim to begin your build during a stretch of dry weather.

CHAPTER SUMMARY

- In thinking about the design of your project, consider the size and lay of your land including all boundary lines, the direction of the sunrise, any slope in your land and how much yard space you want.

- There is literally no end to the number of ways you can design your house. In addition to magazines, TV and the Internet, you can look for inspiration from your own life. Think about what you love in other homes you've seen, how you plan to use each space and your storage needs.

- There are plenty of pros and cons for including high ceilings. Consider the best way to utilise this popular design feature.

- Think about what rooms you may want to include in your home, such as lounge room, kitchen, dining room, play/rumpus room, media room, bedrooms, bathrooms, hallways, study, garage, granny flat, deck/pergola, cellar or other outdoor structures.

- Can't visualise the size of a room? Get a feel for those dimensions by walking around another room of a similar size.

- The design stage is when you need to consider the outdoor structures you want to include because the construction of these structures can affect the construction of your house and vice versa.

- As you draw up your plans, remember to stay flexible – it may turn out that your ideas are not structurally or financially feasible. Don't get your heart set on something until you talk to the professionals. When drawing up your own rough plans, you can make sketches, use graph paper or take advantage of online designing programs.

It's okay to look at other people's designs – such as those of a volume builder – but make sure you don't copy those plans.

There are lots of elements you can add to make your house more energy efficient including double-glazed windows, insulation, proper seals, natural light, water tanks and solar panels.

Different elements go into the classification of the Six Star Energy Rating. Find out the rules for your local area.

You can renovate to include energy-efficient elements into your house but if you can, include them at the time of the build.

The best time of year to plan to do your build is when it's the driest.

WHAT COMES NEXT

At this point, you should have a pretty good idea of what you want to build, including the orientation of your house, the general floor plan, the type of foundation you'd like to use and the type of building materials you'd like to use. Before you find an architect and an engineer to finalise your plans, though, you'll need to figure out how much money you can spend and get your documents in order.

CHAPTER FOUR

ORGANISING YOUR FINANCING AND DOCUMENTATION

By this stage, you will have spent a lot of time thinking about what you want to build. Before you go any further, you'll need to find out just how much money you'll have available for your project. You'll also need to go about getting together all of those essential documents that will protect you legally as you go through the process.

A Word to the Wise

Check the facts

As I said in Chapter Two, the guidelines I've provided here are simply that – guidelines. The rules and procedures will vary widely from one place to another. You really must get expert advice from people who know the laws and procedures that are specific to your area, your lender and your situation. What I've provided here is only meant to give you a starting point.

FINANCING

Perhaps the most vital step of the entire building procedure is making sure you have enough money available to you before you begin your project. Your budget will determine which elements you can include in your design, which materials you can use in your build and how much time you will have to finish the project. If you get this step wrong, you may find yourself in all sorts of trouble.

To begin with, it's important to understand that the type of loan you get is different when you build than it is when you're buying an established home – and the rules for owner builders are different to those that apply when you hire a builder to do the job for you. Let's break down the process so that you'll have a good idea of what to expect.

The construction loan

When you go to buy a house, you can usually expect to pay a small down payment and then borrow the rest of the money to buy the house. The amount of the mortgage is usually based on your credit standing and income. When you build a house, however, you'll usually start with a special type of financing called a **construction loan**.

If you were to hire a builder, your lender would calculate the total amount your project is going to cost. The lender will divide the total amount of your construction loan into four or five separate payments called **progress draws**. By the end, all of the money will have been passed along to the builder.

With a **construction loan**, your lender calculates the total amount your project is going to cost then divides that figure into four or five separate payments called **progress draws** which would be transferred directly to your builder in chunks as the project progresses.

Most of the time with a construction loan, you would only need to repay the interest on the amounts that have been paid to the builder until the project is finished. At that point, your construction loan would become very similar to a regular mortgage and you would start repaying both the principal of the loan and the interest.

The owner builder mortgage

A loan for an owner builder is a little different than the type of construction loan you'd get if you weren't doing it yourself. If you're doing it yourself, you'll get an **owner builder mortgage**.

The difference stems from the fact that most lenders consider owner builders to be high risk. Why? When the people at the bank make a loan, they want to make sure they get their money back. In the case of a standard mortgage, if you don't repay your loan, the bank can take your house. When they're dealing with a volume builder or a licensed independent builder, the banks can feel reasonably confident they'll still have a house to use as security against the loan.

With an **owner builder mortgage**, your lender calculates the percentage of your build you can borrow against based on the value of the lot and quotes from your tradesmen then periodically transfers the progress draws directly to you.

As an owner builder – someone who has probably never built a house before and who isn't a licensed builder – you can't provide the same level of reassurance to the lender that your loan will be repaid or that they'll have collateral to take away in the event that you don't pay.

Drawbacks of an owner builder mortgage

To offset the lack of security, the majority of lenders who are willing to do business with owner builders will usually only lend you 60 per cent of the total value of your land and construction cost (although a couple will go as high as 80 per cent). In fact, lenders rarely even consider what your house is going to be worth once it's finished.

Instead, they look at the value of the empty lot first. Then they'll estimate a TBE – that is, 'to be erected' – valuation based on the amount of money you plan to spend on labour and materials. These figures will usually be based on quotes you provide from your tradesmen (although sometimes lenders will consider estimates).

Your lender may (or may not) require you to pay a hefty down payment or own your plot of land outright before you start building. You may also be expected to pay as much as 40 per cent of the total cost yourself before you can start drawing on your loan amount.

Benefits of an owner builder mortgage

Though you may not be able to borrow as much as you might if you were hiring someone else to do the work, you get to have a lot more control over how your money gets spent.

As the owner builder, you'll still have your loan divided into chunks – the progress draws – but you set up how much money you'll get with each one and how often. Also, the money usually goes directly to you instead of to a builder, so you know exactly how much will go towards labour and materials instead of leaving those huge amounts in the hands of someone else.

At the end of the project, you'll still have to pay back the loan, but you are very likely to discover that – if you've done the job properly – the value of your home may far outweigh the amount you owe. And that's a good position to be in.

What to consider before you borrow

Whether you're building or buying, taking on a new home loan is always going to be a big commitment. In thinking about organising your finance, there are a lot of elements to consider.

Do your homework

Depending on where you live, there will be different laws governing what owner builders can and cannot do in terms of financing. Laws are constantly changing, and a small change in your state's regulations can make a big difference to you as an owner builder. Be sure to discuss all the fine print with your lender before you get started so that you'll know what to expect.

The best place to begin will be to research all the lenders you might possibly use. In most cases, you'll probably only have a few options available to you because not every bank will lend to an owner builder. Each lender will have their own procedures in place so find out what they will require of you, what their terms are and which lender will be the best fit for your situation.

It would be worth your time to make an appointment to visit all the potential lenders and ask a lot of questions, including what interest rates they charge, the length of loan you can get, any time restraints that may be posed, penalties you may have for early repayment, your options in terms of refinancing your loan and the possibility of borrowing more money later if you need it.

If you don't understand the process, sit down with someone who does and learn as much as you can.

Consider if you qualify

Before you begin the process of securing financing, keep in mind that your lender will want to make sure you're a viable borrower. Criteria will change from one lender to the next, but be prepared to provide all the standard information including your salary, your assets, your expenses, your debts and your credit history.

If you've been a credit risk in the past, there's a good chance you'll have a very hard time qualifying for an owner builder mortgage.

A Word to the Wise

Get more money up front

Try to structure your payments or progress draws to suit your needs. If I had it to do all over again, I would have set up bigger loan payments earlier in the build and had a smaller final disbursement.

If you don't schedule those payments properly, you may end up running out of money before the next disbursement comes which can hold up construction and cause all kinds of problems. Before you finalise your financing, you should be able to draw up a solid schedule and budget and let your foreman help you schedule those progress draws most effectively.

Get pre-approval

At this stage, you probably won't be able to settle on a specific loan amount. You'll be able to formalise your budget after you've met with your foreman and you've obtained quotes from your tradesmen, and your lender will want to see all of these quotes before they advance any money to you.

All you need to do for now is find out the maximum amount you can comfortably borrow from each potential lender. With that figure in mind, you will be in a better position to choose materials, designs and building techniques when the time comes to set those decisions in stone.

Over borrow

If you don't borrow enough money, you may find yourself in a very sticky situation. If you run out of money and something goes wrong, your project may drag on years longer than you'd intended or you may end up having to cover a shortfall of tens of thousands of dollars. Twenty thousand dollars may not sound like much when you're talking about a project worth a few hundred thousand dollars, but you may find it's tougher than you thought to come up with that money late in your project.

After you've settled on a specific yet flexible budget, you'll be able to come back to the lender that's right for you with all of your quotes and sign the documents formalising your loan. Regardless of your budget, I would highly recommend you borrow more than what you think you'll need.

Why? Any number of things can go wrong with a build – jobs sometimes get delayed, building materials sometimes increase in price, tradesmen sometimes need more time than they originally anticipated and the weather might suddenly turn against you. You need to be prepared to withstand those potentially expensive problems.

Also, depending on your lender and where you live, you may only be able to finance once. And in terms of construction, it's not usually a good idea to build a house one element at a time, leaving the structure exposed to the weather for long periods of time between each phase of the project.

As a rule of thumb, it's better to organise your loan up front. Plan to borrow twenty to thirty per cent more than what you think you'll need. If you end up with too much money at the end of the project, you're free to pay it back towards the loan and save yourself some interest.

If you can't borrow more than what you think you need, I would highly recommend you not embark on this project at all. Otherwise, it may take you years to finish your house and the stages you have already built could be ruined by the time you have the money you need to continue.

Consider non-building related expenses

Where will you be living while your house is being built? If the project goes over schedule, then you may end up paying for a separate place to live in while the build project happens. If you do need to rent for a period of time, the time for your building project needs to be the same time frame as your rental period so that you don't end up paying double for accommodation.

You may also need to consider other non-building related costs such as power and water for your tradesmen to use during the build. And keep in mind that any time you move house, you'll always incur extra costs such as utilities reconnection fees.

Is it worth it?

There's a difference between the amount of money a lender will let you borrow and the amount you're really comfortable repaying. You don't want to be left short during the build, but if you end up mortgaging the maximum amount of money for which you qualify, you may end up with a hefty monthly repayment – at that point, 30 years becomes a very long time. Think about just how much that monthly bill is likely to pinch when choosing your materials and other expenditures.

A Word to the Wise

Over borrow

Delays can happen to anyone. When I built my house, a few different things didn't go the way I'd planned. I borrowed exactly the amount I thought I was going to need, but at one stage, I ended up short by several thousand dollars and I wasn't able to get more funds when I needed them.

I didn't put enough time into planning my overall figure and my build ended up taking a lot longer than what I'd originally intended. Just be careful to consider every cost and you can avoid making the same mistake I did.

One big loan or pay as you go?

Depending on your financial situation, it may be tempting to avoid taking out a large mortgage and build your house or do your renovations one step at a time. Paying as you go might be possible – especially for an interior renovation – but in general, it's not a good idea.

If you're building or rebuilding an entire house, or if your renovations will include taking down exterior walls, I would recommend not trying to build one step at a time. One good reason is that, once you let your tradesmen go, you may have trouble getting them back. Tradesmen are quite often booked for eight or ten weeks in advance. If you're going to try building one step at a time, you'll want to make sure you've got a builder who can cater to your needs.

Also, when a job drags on, you run the risk of materials being exposed to weather - not ideal and not recommended. And if you can't pay cash as you go, getting several smaller loans one at a time may end up being more expensive in the long run.

If you're doing a renovation in which your house will remain locked up throughout the build - while you redo one room at a time - then you may be able to pay as you go without taking out a loan.

DOCUMENTATION

By now, you should have a good idea about the design you'd like to build, the materials and features you'd like to include and how much money you can spend. Now would be a good time to start getting your documents in order.

Permits, contracts and licenses

The rules for owner builders vary from state to state and will depend in part on what type of build you want. I can't give you a definitive answer here as to exactly what you need because every case and every council is different but there are a few licences, permits and insurances you will almost certainly need either before you begin your project or during the process.

Probably the best place to begin is your state government's website. The following is an excerpt from the Victorian Building Authority website.

> An **owner builder** is someone who takes on most of the responsibility for domestic building work carried out on their land.
>
> As an owner builder you would need to obtain building permits, supervise or undertake the building work, and ensure the work meets building regulations and standards. In Victoria, an owner builder can only build or renovate one house every three years, and must intend to live in the house once completed.
>
> If you wish to be an owner builder, and the value of the proposed building work is more than $12,000 (including labour costs and materials) you will need to apply for a certificate of consent.
>
> **Please note: you and a builder, contractor or tradesperson must enter into a written contract for domestic building work costing more than $5000.**
>
> The certificate must be provided to your building surveyor in order to obtain a building permit.
>
> As an owner builder, you will be given the same level of protection afforded to other consumers under the *Domestic Building Contracts Act 1995.*

The website goes on to explain that: 'If you engage a number of builders to construct various parts of the work and also do some small components yourself, you ARE an owner builder.'

The documentation you'll need will depend on a number of factors including where, what and how you build but the procedures are generally very straightforward and not very expensive. It's important that you know about everything you will need and get it arranged in a timely manner – otherwise you may see delays – and when you're building, time truly is money.

Don't be intimidated by the process. It may sound like a lot to do, but it's really just a matter of researching the requirements for your particular state or council.

Insurance

Any number of things could go wrong when you're building. Even for a small renovation, you need to make sure everyone on the job site is covered. For this reason, you make sure you are properly insured. Again, exactly which insurance you will need will be specific to your build, so find out from your council. These are the most common types.

Public liability insurance

As an owner builder, you need to arrange **public liability insurance**, which protects you if anything happens to the public as a result of your build. You might have a truck come in that rips down a power line and kills three people. Though we would hope that would never happen, your public liability insurance will cover you for $20 million dollars in case it does.

Public liability insurance covers you if someone in the public is injured as a result of your build.

Construction insurance covers your property in the event of vandalism, theft, inclement weather, fire or other catastrophes – it can even cover tools and personal injury.

Construction insurance

Like with any house, you need to make sure your property is covered in the event vandalism, theft, inclement weather, fire or other catastrophes that may happen while you're building. This is what **construction insurance** is for.

Depending on which company you use, you can even get cover for tools and personal injury. Shop around and see what's available for the best price in your area.

Domestic building insurance

Remember when we talked about legal issues you need to consider back in Chapter Two? In some cases, you as the owner builder need to be able to guarantee the workmanship of your build even after you sell the property. This insurance is called domestic building insurance.

> **Domestic building insurance** covers the person who buys your home in the event that you die or go bankrupt and your house is found to be incomplete or defective. This insurance may be required depending on where you live, so find out.

> **Workman's insurance** – also called WorkCover – protects the people who work for you in the event that one of your tradesmen gets hurt while working on your hose. It covers benefits such as income replacement, medical treatment, rehab, legal costs and even lump sum compensation if the injury is serious.

Workman's insurance

Most states require employers to have insurance that will protect their workers in the event of injury, both big and small. This type of **workman's insurance** is usually called WorkCover.

If one of your tradesmen gets hurt while working on your house, they're entitled by law to benefits such as income replacement, medical treatment, rehab, legal

costs and even lump sum compensation if the injury is serious.

If you're paying salaries to tradesmen, you need to make sure you're able to look after them in case of accidents.

Income protection

You've made sure the public is protected and your tradesmen are looked after – but what if something happens to you during your build and you can't pay?

There's never a good time to get injured or lose your job, but in the middle of building your house is particularly bad. **Income protection** will make sure you are looked after if something happens and you stop earning your regular income.

The White Card

Everyone who does construction work in Australia is required to have a **White Card**, which is the common term for General Safety Induction Training. To get your White Card, you need to attend a training course that will provide you with information about the Work Health and Safety (WHS) laws that apply to construction work and ways you can work safely.

A White Card isn't technically insurance in the sense that you're not buying a policy, but before you go onto a job site - even your own - you need to know what risks to anticipate and how to avoid them. And if you move states, your White Card is recognised nationwide.

There are online courses that take approximately four hours, cost under $100 and you get a certificate within 24 hours. Search 'white card' on the Internet for more information.

Income protection insurance will provide you with money if something happens and you stop earning your regular income.

The **White Card** is required for everyone who does construction work in Australia – it is the common term for General Safety Induction Training.

A Word to the Wise

Keep your tradespeople safe

Every time there's a serious injury, there will be an investigation. If you're sending someone to hospital with a huge laceration or some other major problem, you'll find that WorkCover is likely to investigate your property to make sure you've had all the correct safety procedures in place to prevent injuries. If they find you were negligent by failing to set up a safe worksite, you will be liable. You may be required to pay huge fines and you could even go to jail.

THE OWNER BUILDER COURSE

If you're really keen and you think you could use some extra help, it's not a bad idea to consider taking an owner builder course. There are plenty of courses available online to prepare you in more detail for what to expect at every stage along the way.

These courses are generally specific to each state, so by taking a course, you can feel confident that you know all the relevant rules that apply to your particular area. Some of these courses even include White Card training so that you can get both at the same time. In most cases, you should be able to complete an online course in a weekend, including all the study and assessments.

CHAPTER SUMMARY

- It is absolutely essential that you get your financing right - if you run out of money in the middle of your build, you could end up in real trouble.

- The type of loan you'll get as an owner builder is different from the type you'd get if you were simply buying a house or hiring a registered builder to do the job for you.

- With a construction loan, your lender divides your loan into payments called progress draws and disburses them to your builder periodically throughout the project.

- With an owner builder mortgage, your loan amount is based on your quotes and the progress draws are usually paid directly to you.

- With an owner builder loan, you will be limited as to the number of lenders that are available to you, and you probably won't be able to mortgage the full amount of your project - 60 per cent is usually standard or 80 per cent in rare occasions.

- Because you won't be able to borrow 100 per cent, you need to be prepared to pay between twenty and forty per cent of the project costs from another source other than your loan.

- As with any construction loan, you won't get the entire sum of your loan in one chunk up front - instead, the loan will be sent in portions at various intervals throughout the project. As an owner builder, though, you will get to decide how often your lender sends you money and exactly how that money is spent.

Before you choose a lender, you should research your options and consider whether or not you're likely to qualify for the loan. If you are a credit risk, you almost certainly won't be able to get financing as an owner builder.

You won't be able to finalise your loan until you have quotes in hand for materials and labour. Start by getting pre-approval so you'll know how much you have to spend.

You should definitely over borrow. I cannot stress enough how difficult your project will become if you run out of money. Consider non-building related costs and whether or not you really want to be committed to the maximum amount for which you qualify.

It's generally not a good idea to get several smaller loans and draw out your project over a longer period of time. It's much better to get one big loan and get the project finished as quickly as possible.

The laws regarding which licenses, permits and contracts you need vary from one state to the next. Do your homework so that you'll know for sure what you need.

You must be properly insured before you begin your project. You may need to organise public liability insurance to protect the public, construction insurance to protect your build, domestic building insurance to protect the people who buy your house, workman's insurance to protect your tradesmen, income protection to protect yourself and a White Card to make sure you know how to work safely.

A good way to make sure you're completely informed is to take an owner builder course that's relevant to your particular area.

WHAT COMES NEXT

You've got a lot of great ideas just itching to come to life, and now you know exactly how much money you'll have to make it happen. With your documents all lined up, you're ready to get your rough sketches finalised into the blueprints that you and your foreman will follow to build the house. Time to find a land surveyor, an architect and an engineer.

CHAPTER FIVE

ORGANISING YOUR LAND SURVEYOR, ARCHITECT AND ENGINEER

Now that you know the maximum budget you're comfortable working with, it's time to get your rough plans drawn up by a professional. First, you'll need to find a land surveyor, then an architect and an engineer.

THE LAND SURVEYOR

Before you can determine how big the perimeter of your house can be, you need to be absolutely certain of the boundaries of your entire property - both side to side and top to bottom. The person who does this job is the **land surveyor**.

A **land surveyor** accurately measures the exact boundaries of your property.

What does the land surveyor do?

A qualified land surveyor is able to accurately measure the exact boundaries of your property. Using a lot of really fancy sounding equipment like advanced GPS, Robotic Total Stations and aerial and terrestrial scanners, he or she can take detailed measurements and run them through specialised software to come up with a comprehensive report.

Your land surveyor will also do the other job you probably associate with this work - putting pegs with orange flags in the ground to mark your land. In addition to measuring your property, the land surveyor will make sure you're following all the guidelines required by your local council.

Just doing a renovation? If you're expanding your home, you probably still need to have a land surveyor come and check out your boundary lines. Whether you're building up or out, you don't want to build beyond the envelope of your property - if you do, you may find yourself in a dispute with your neighbours that you probably won't win.

What is your building envelope?

One of those guidelines set forth by your council will be how high you can build your house. In this context, the **envelope** is the three-dimensional zone into which your entire property needs to fit.

Your local council will determine not only the length and width of the land but also the maximum height to which you can build without obstructing the view of your neighbours. You'll definitely want this information if you're planning a renovation or you've been considering adding a second storey. The rules are different depending on where you live, so make sure you find out.

> The **envelope** is the three dimensional zone into which your entire property needs to fit, including length, width and height.

If you don't get your boundaries measured properly or you hire someone without the proper qualifications, you may end up building your house too high or in the wrong spot. The last thing you want is to have to pull down your house because you're just a few metres too close to the borderline.

What should you look for in a land surveyor?

Experience is always a good indicator of a land surveyor you can trust. Only a licensed surveyor who holds a practicing certificate is qualified to provide you with an authorised survey. He or she should provide you with both a written quote prior to conducting your survey and a signed certificate indicating the job has been done properly within 30 days of completing the survey.

All qualified land surveyors should hold public liability and professional indemnity insurance.

Where can you find a good land surveyor?

You can start by asking around – word of mouth is always a good bet.

As for other services, you can find plenty of land surveyors advertising online, but you can confirm his or her qualifications by checking with your local land surveyors licensing board.

Q&A

I'm sure my land has already been surveyed. Do I really need to get it done again?

If you're buying an established home, you may be able to get a report of the exact boundaries of your property from your local land titles office or council. Depending on what you want to do, you could possibility skip the step of having your land surveyed but probably not. Ask at your local council.

If you have your land surveyed, you may find that the existing fence is in the wrong place – if the fence is cutting into your property, you have the right to ask for that fence to be moved.

THE ARCHITECT

Once you have the results of your land survey, you can take your rough plans to an **architect.** The architect is the person who will take your ideas and turn them into official drawings that will result in a plan for building your house that meets code and adheres to all laws and regulations.

An **architect** designs your entire environment, including all indoor and outdoor spaces, maximising the efficiency and appearance of the space you have.

What does the architect do?

The short answer is that he or she turns your rough plans into official blueprints based on what you want, but there's more to the job than just that. Part of an architect's job is to know all the rules and regulations your local council will require and how to apply those rules to your house.

He or she can also give you a rough idea of how well your budget matches your designs. You won't be able to come up with a specific number of costs until you organise your tradesmen. Your architect can, however, give you an estimate of the overall cost you can expect and suggest materials and techniques to help you stay within your price range. We'll talk more about drawing up a specific budget in Chapter Seven.

It's a very good idea for you to listen to your architect's ideas and suggestions. Rather than looking at simply the outlines of the house, your architect will look at the property as a whole and make suggestions for the best ways to use your entire space, both inside and out.

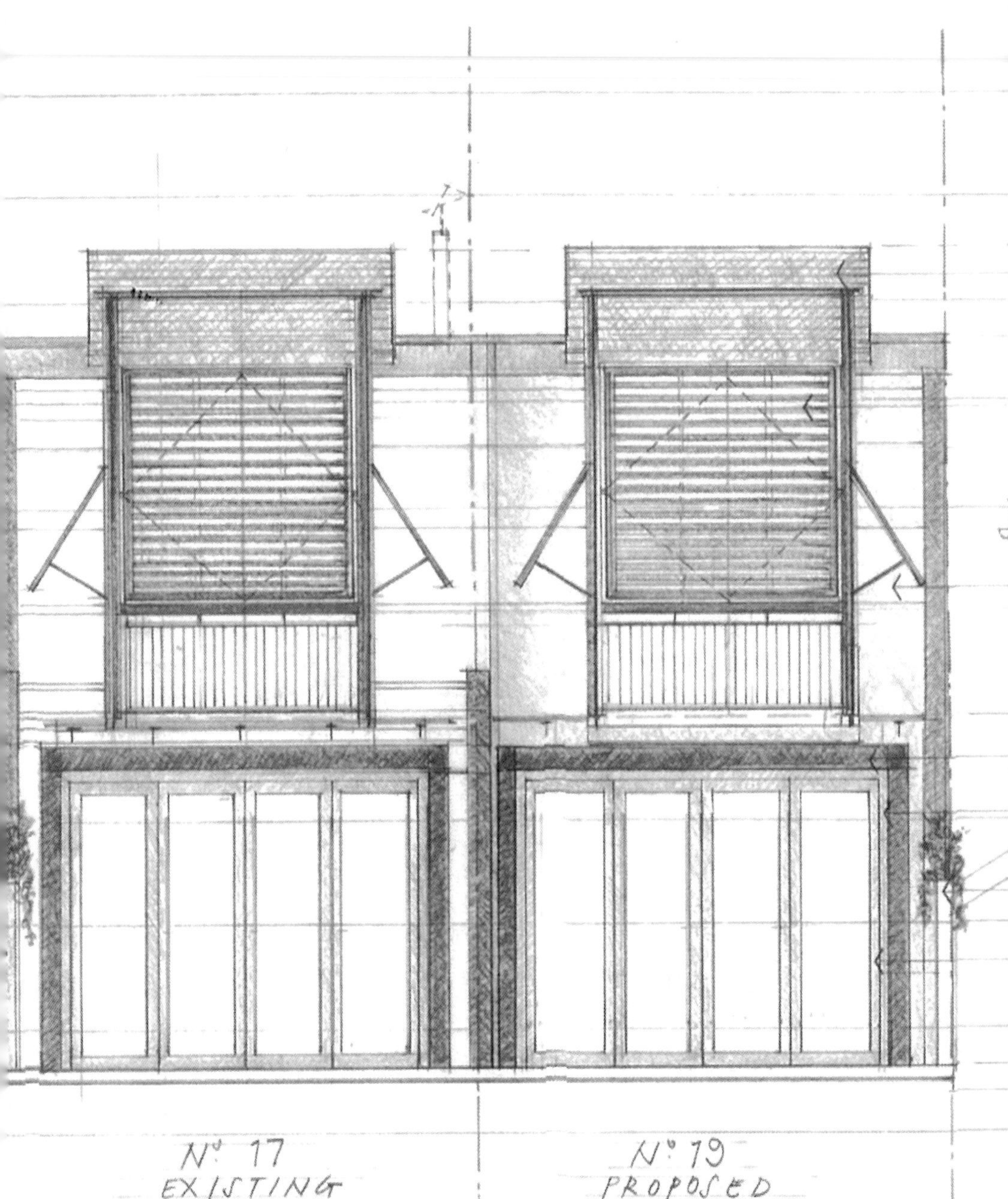

NOTE: REAR BUILDING LINE
SET-BACK 850 MM.
FROM REAR BUILDING LINE
ESTABLISHED @ Nº 17

A good architect is a creative professional, which means he or she is likely to have a lot of great ideas in terms of design, techniques and procedures that might not have occurred to you before. More than just creating the best look for your home, your architect can also tell you things such as the best way to absorb or reflect sunlight and how you can maximise cross-flow ventilation.

Most importantly, your architect should listen to you and do his or her best to incorporate the elements you want in your house at a price that fits your budget. It will be your job to come to your architect prepared with as much information as you can – the more specific you can be, the better.

What should you look for in an architect?

To become qualified, an architect must have studied for at least five years at an accredited university. Also, he or she will have had to pass a registration exam and obtain the required number of hours of practical experience. Your architect should be able to produce documentation of his or her qualifications. He or she should also hold professional indemnity insurance. As with a land surveyor, an architect with several years of experience is ideal.

Where can you find a good architect?

A quick Google search will turn up plenty of architects in your area, but word of mouth is probably the best.

A lot of people choose their architect based mostly on price, but there are plenty of other factors to consider. Get a lot of quotes – you can ask for hourly rates or get a fixed quote. Also, ask for references and look at previous work they've done.

If you want to make sure your architect is properly qualified, you can check to see if he or she registered or a member of the Australian Institute of Architects.

Draftsman or architect?

A lot of people prefer to hire a **draftsman** to draw up their plans rather than an architect. So what's the difference between the two? Draftsmen are basically trained to create detailed technical drawings including precise specifications, generally using design software, often based on an architect's plans.

A **draftsman** creates the technical drawings based on the architect's designs.

The architect designs the entire environment, taking into account both appearance and efficiency plus building and council regulations. The job of a draftsman is much more specific – they only specialise in producing the blueprints from the architect's designs. In fact, most architects work alongside draftsmen to come up with the final plans for a building.

The benefit of using a draftsman is that, because they only create one element of the design, they tend to charge a lot less money. At the same time, a draftsman doesn't have the same qualification as an architect. Unlike architects, who must complete a university degree as part of their accreditation, draftsmen study to earn a Certificate at a TAFE college.

Whether you want to hire an architect or a draftsman depends in part on how confident you are in your own abilities. If you're building an entire house, especially a high-end one, you'll probably want to hire an architect.

You would also do better to use an architect if you're not sure of the best ways to maximise the space you have.

If you're only doing some renovations, however, or you've done a lot of research and you think you've taken advantage of every opportunity for design and efficiency in your build, a draftsman might be a better bet.

A Word to the Wise

To every rule …

Every council will have certain codes of practice in place about what can and can't be built. If you really want something that is outside council's code of practice, you may be able to get your local council to make an exception.

In general, if your designs are contrary to your council's regulations, you can apply for a dispensation – there isn't an exception to every rule, but it doesn't hurt to try.

Getting the most out of your architect

You can expect your architect to be very professional but there are a few ways to make sure you maximise the time you spend with him or her.

Speak up

You should have a very good idea of what you want going into your meetings with the architect. Now is not the time to be indecisive, although it is wise to have a few different options in mind.

If something doesn't look right, say so but remember that there's a right way and a wrong way to speak to people - you'll accomplish nothing by being hostile. Be prepared to discuss all the available options with your architect. The more informed you are, the better, so do your research.

Listen

While you want to be able to express your ideas, your architect should know what he or she is talking about and it's his or her job to act on your behalf. Good architects do everything they can to bring your vision to life, but at the same time, it's a good idea to listen to their suggestions. If your architect proposes something you absolutely cannot live with, ask him or her to help you find a way to solve the dilemma.

Ask questions

Before you get started, you shouldn't hesitate to ask as many questions as you need to. If your architect tells you that some element of your proposed design isn't possible, you are entitled to ask for an explanation. If anything your architect proposes doesn't make sense to you, ask for clarification. And if there's any confusion regarding your contract with your architect, ask for verification.

Document

Architects can cost a lot of money so you need to document all of your meetings. If you're suggesting changes and the architect isn't making them, he or she may try to charge you wrongly for an amendment.

You really have to take minutes or record your meetings with architects because they may inadvertently charge you for making changes even if they've made the mistake.

Q&A

Which came first, the architect or the foreman?

One of the most vital people you'll work with on your project is your foreman. You're probably better off getting your plans in order with the architect before you go to your foreman.

If you don't feel comfortable meeting with the architect on your own, however, you can pay your foreman an hourly rate to discuss your plans with you prior to meeting with your architect. You can even have the foreman sit in on your meetings with the architect. We'll talk more about hiring a foreman in Chapter Six.

Making changes after your plans have been drawn

You'll spend the bulk of your time with your architect long before the build happens. As the house starts to go up, however, you may discover there are changes you want to make.

If you want to make a change in your plans, you'll need to get your architect to include them. What might look like a small alteration to you may have far-reaching consequences in terms of foundation and framing, so you'll want to make sure your changes are feasible.

As soon as you decide to make a change, you'll need to go back to your architect and have him or her amend the original designs. Your plans have to be exactly as you build so that your building inspector can compare your plans to the work that's been done.

If you make a change without amending your plans, you probably won't pass inspection – which means you won't be able to move on to the next step.

The good news is that, as an owner builder, your architect and engineer should only charge you a few hundred dollars to make these changes – a large volume builder will ask for thousands, even if the change is a small, simple one. If you do ask for changes, find out the cost and make sure your architect includes them as sometimes those requests can be overlooked.

THE ENGINEER

Now that you've had the land surveyed and you have the architect's plans, you'll need to see the **engineer** whose job is quite detailed and will encompass a range of elements the average person would probably never consider prior to building or renovating.

The **engineer** is the person who works out the science and maths involved in turning your designs into a reality.

What does the engineer do?

An engineer will take the architect's designs (which will include the blueprints that have probably been drawn up in detail by a draftsman) and work out all the maths and science necessary to make those designs possible. We'll talk more about what his or her role involves in just a minute.

What should you look for in an engineer?

There are different levels of qualification for engineers, including Professional Engineer, Engineering Technologist and Engineering Associate. Any of these qualifications will require completion of an engineering technology course, either from a university or a specialist college. Your engineer may be a member of a professional organisation such as Engineers Australia and, like with an architect, he or she should carry professional indemnity insurance. And the more experience, the better.

Where can you find a good engineer?

Your architect and your engineer will work together closely, so it's always good if you can use an engineer that your architect has recommended. Though it's not likely to happen, if your architect can't provide you with a recommendation you can look online but ask for references. And, of course, make sure you have a written agreement for what work the engineer will do for you and at what cost.

The role of the engineer

Your engineer has to consider a long list of factors in determining the best way to bring your ideas to life.

The foundation inspection

If you've bought an established house and you're planning to renovate, you'll need to make sure the foundation is in good shape. To determine the condition of your foundation, you'll need to have someone – a carpenter or other tradesman – come and dig down to the bottom of your foundation, either to the bottom of your concrete slab or to the bottom of a few of your stumps. This person will take photos to determine the depth at each corner of the house.

The foundation inspection will determine whether or not the depth of your foundation is adequate, if there's been enough concrete used, if the stumps are rotten or if there's a lot of cracking in the foundation.

Based on this inspection, you can decide whether or not you'll try to keep the foundation.

This step is equally as important if you're doing renovations. If you're removing a wall, for example, you'll need to make sure your foundations are strong enough to handle the redistribution of weight in your frame.

The soil report

After the foundation inspection, you'll need to get a geotechnical expert to come to your property and conduct a soil report. This report will determine how much the land on your plot is likely to shift or change once you've built.

There are six classifications into which your soil is likely to fit as shown in the following table.

Soil Class	What it means
A	the most stable and should experience little or no movement
S	may experience a little movement
M	can see moderate movement
H	high levels of movement
E	extreme levels of movement
P	problem sites as a result of mine subsistence, landslip, coastal dunes or soft soils with a lack of suitable bearing

Based on the soil report, the engineer will make some calculations and recommendations. In general, the less stable your soil is, the more sturdy your foundation and frame will need to be – and more sturdy usually means more expensive.

Your engineer is likely to be in touch with a geotechnical expert so you probably won't have to engage that person yourself. If you do, just do a search for a geotechnical specialist – all the same rules of hiring apply, including a check for qualifications, written quotes, references and experience.

The termite report

The engineer will provide you with the vital termite report and let you know whether or not you need to invest in termite preventatives.

Termites are perhaps the biggest enemy of the builder. If they get into your structure, they can eat your entire house and you won't know until it's too late. For this reason, I personally recommend you protect against termites even if you're not in a high-risk area.

To protect against termites, your engineer may suggest you use steel frames such as a Supaloc steel frame or you can build your frame with bricks. There are also new framing products called **LVLs** that have chemicals built into the timber, which prevent termites – termites simply won't eat LVLs.

LVL stands for laminated veneer lumber.
Great protection against termites – made from two pieces of 90x45mm wood that have been laminated together to create a thickness of 90x90mm.
Termites simply won't touch them.

Climate considerations

Your engineer will also provide you with information relevant to building in your particular region. If you live in Western Australia, for example, you can nearly count on needing to protect against termites.

Your engineer will tell you if you need to build a house capable of coping with extreme humidity, cyclones, snow or sea air – and he or she can also recommend the best materials available for your region.

Cost considerations

Like your architect, your engineer should be able to give you an idea of the costs of different materials and building techniques and make recommendations best suited to your situation.

Consult the engineer on the costs involved and also the pros and cons of different techniques. Every council and every area will require certain building techniques, and it's up to both the engineer and the architect to work to those methods and laws.

What are some pre-building expenses I can expect?

For obvious reasons, your land surveyor, architect and engineer will all begin their work prior to the construction of your house, and at this stage you probably won't have your finance finalised. Be prepared to have some funds set aside to cover the costs of getting your boundaries measured, having your plans and blueprints drawn up and getting the necessary testing done for the engineering of your project.

A Word to the Wise

Beware of old houses

In the old days (when I first started building and before), they often didn't have soil reports and the ones they did have weren't as good as the ones we have now. I've seen houses that didn't have a foundation at all and were simply built directly on top of the ground.

If you're dealing with an older house, make sure you know all its flaws – a home built twenty or thirty years ago was almost certainly not built to a standard that comes close to houses that are being built today.

It's worth the money to get any house closely inspected before you buy so that you know for sure what you're dealing with before you make any further decisions. We'll talk more about finding a building inspector in Chapter Seven.

CHAPTER SUMMARY

- With your own designs in hand and pre-approval for financing, you're ready to turn those rough plans into proper blueprints from which your house will be build. At this point, you need to engage a land surveyor, an architect and an engineer.

- The first person to see your property should be a land surveyor who will determine the perimeter of your site on all sides and advise on how high you can build, and what to take into account in terms of council regulations. Without a qualified land surveyor's report, you won't have much legal ground to stand on in the event of a dispute.

- In hiring a land surveyor, look at qualifications and experience in addition to price. Also, make sure the surveyor is properly insured. Look for a land surveyor by word of mouth or online, and check the person's qualifications with your local land surveyors licensing board.

- Even if you are able to obtain boundary lines for your property from your local title office or council, you'll probably still want to have your land surveyed before you build or renovate.

- Next comes the architect, who will consider the entire environment of your build. Part of his or her job is to have blueprints drawn up by a draftsman, but the architect will also make lots of suggestions to maximise your space and advise on the various options that are available to you.

- As with a land surveyor, your architect should have experience to add to his or her university degree plus adequate professional insurance. Be sure to draw up a written contract with your architect to clarify the work and the price.

You can find an architect by all the usual methods – online, in the paper or by word of mouth. Ask for references and check registration or membership of the Australian Institute of Architects.

In some cases, you may want to save some money by hiring a draftsman directly to do just your drawings rather than hiring a more expensive architect to design the entire property. Because this job is more specialised and less comprehensive, a draftsman will have a TAFE certification rather than a university degree.

There are some cases in which you can get a dispensation to bypass certain council regulations. It doesn't hurt to ask.

There are lots of ways to get the most out of your time with your architect. Be sure to speak up with lots of specific ideas, listen to suggestions, ask lots of questions and document every session you have with him or her, both before and during the build.

If you don't feel comfortable meeting with your architect on your own, you may want to hire your foreman first and ask him or her to attend meetings with you. In general, though, you'll have an easier time showing the foreman what you want if you already have your architect's plans in hand.

If you want to make changes to your plans at any stage in the build, make sure your architect includes those amendments to your plans. Otherwise, your alterations may not be structurally feasible. More importantly, you may not pass inspection if your finished project doesn't match your plans.

From the architect, you'll go to the engineer. Your engineer will calculate all the maths and science necessary to make your blueprints a reality. All the same guidelines apply to finding and hiring an engineer as apply to finding and hiring a land surveyor and an architect (although it's a good idea to use one your architect recommends).

Working from your architect's plans and within the boundaries of your land surveyor's report, your engineer will consider the foundation inspection, the soil report, the termite report, climate considerations and cost considerations.

You'll need to hire your land surveyor, architect and engineer before you begin building, so their services will not be covered by the financing you'll get later from your lender. Make sure you're prepared for these and other pre-building expenses.

Be especially careful to get older houses thoroughly inspected before making any decisions about renovating or rebuilding as those old houses probably haven't been built to today's standards.

WHAT COMES NEXT

With firm plans in hand, you're ready to find the person with whom you will work very closely throughout the entire build - your carpenter/foreman. This person will be vital to your project as he or she will help you oversee the construction of your project from the ground up, so choose wisely.

CHAPTER SIX

ORGANISING YOUR CARPENTER/ FOREMAN

Now that you have your plans from the architect and the results from the engineer, you can start looking for a foreman. This person will do the same things for you that I do for the contestants on *The Block*. The quality of your foreman is likely to make or break your project, so there are plenty of factors to consider when hiring him or her.

In general, a carpenter is involved with the construction of a house from beginning to end. A good carpenter can perform a foundation inspection, make sense of the plans, build the subfloor over stumps, complete the framing of the house and do a lot of the jobs involved in the lock-up and fit-out stages. Your carpenter can even build your decks and some of your other outdoor structures.

Because you'll need a good carpenter throughout the project, it makes sense to hire a competent carpenter to be your foreman. That way, you cover both roles at the same time with one person.

The key is to make sure you define every role you'd like your foreman to do prior to starting the project so that both of you are very clear about what you want done and how you will pay for the work.

What's the difference between a carpenter/foreman and a builder?

You may be able to hire an independent builder to act as your carpenter/foreman. In general, a builder is responsible for overseeing all of the construction of an entire project. Most of the liability lies with that person who should be insured accordingly.

In contrast, a carpenter/foreman simply oversees the project but most of the responsibility lies with you, the owner builder. You can hire a builder to be your foreman, but if you do, you need to make it very clear that you are the owner builder. You will also need to hire a separate carpenter.

ROLE OF THE CARPENTER/FOREMAN

Your foreman will have two main jobs - to complete the carpentry work and to help you oversee the entire project - but there are a lot of other roles that your foreman will fill as well.

Help you find tradesmen

A good foreman should be in contact with a number of good tradesmen with whom he or she has worked with in the past. As an owner builder, it's your responsibility to make your own assessment of each tradesman who comes to work on your project, but it's worth listening to your foreman and trying to use the same people he or she trusts. We'll talk more about organising your tradesmen in the next chapter.

Supervise the tradesmen

Once construction begins on your property, your foreman should be there most of the time.

As a professional, your foreman should be able to spot any problems in build quality as they happen and make sure your tradesmen are arriving on time and sticking to the schedule they've agreed to.

Liaise between you and the tradesmen

If there are problems with your tradesmen, it can be part of the foreman's job to act as the liaison between you and those contractors.

If you don't feel confident doing it yourself, your foreman can approach the tradesman, point out flaws or other concerns and ask for an explanation or instruct them to rectify the issue.

Running weekly site meetings is another job the foreman can do for you. We'll talk more about running a site meeting in Chapter Eight.

Liaise between you and the land surveyor, architect or engineer

In most cases, you probably want to have very specific plans drawn up by the planning professionals prior to hiring a foreman.

If you're a little intimidated by the surveyor, architect or engineer, and you don't think you'll be able to get your ideas across, you might want to find a foreman first who can help you communicate with those other professionals, both before and throughout your project.

Help you draw up a realistic budget and schedule

Once you've organised all of your tradesmen, you should be able to create a fairly detailed budget and timeline based on the estimates of your contractors.

It is vital that you stick to your budget and schedule as closely as you can. If you start running late or you spend too much on materials and labour early on, there's an excellent chance you'll run out of money before your project is finished – and that's never good. We'll look at a couple of examples in the next chapter.

Q&A

Does my foreman have to be a carpenter?

In a word, no – if you find someone you think would do a great job as your foreman, feel free to hire that person even if he or she is not a carpenter. It just makes sense to have a carpenter do double duty as foreman because you'll have a carpenter onsite pretty much from beginning to end.

A Word to the Wise

Have fun with it

Other than my own house, I think my favourite building project I've ever been involved in was a renovation I did for a very successful chef. He and his wife started with a four-bedroom weatherboard home on a massive block of land, but the clients wanted to redo the whole house – only about twenty per cent of the original structure was left standing.

I was the foreman on that job and as we were building, the guy would have ideas on the spot. At one stage, he said, 'I think I might add a wine cellar in here.' As we were finishing off the cellar, the other builders and I jokingly suggested that he needed to put in a beer line as well. Half an hour later, we were putting in a beer line.

We noticed that the client hadn't included a track door into his cellar, so we said, 'Why don't we incorporate that into the deck?' We ended up putting in automatic push button doors for the deck and every modern convenience you could think of. As we were building the house, we would throw out ideas of what they could and couldn't have, and a lot of the time, the guy just ran with our ideas.

We started out with a budget for a $1.2 million renovation, but in the end, it turned into a $2.4 or $2.5 million renovation. Obviously the clients knew their original budget and how much more they could afford to spend. With each addition, we let the couple know how much extra it was going to cost to do what they wanted, and they would say, 'Yep, let's do it.' It was great fun.

We had a hard time working with the architect on that job because we would tell him the changes the client wanted but at the site meeting the following week, we would see that the architect hadn't made the changes to the plans. We were able to work around it, but it was difficult and we really had to stay on him. You have to make sure your architect or draftsman understands your vision and provides you with what you want, rather than with what he or she wants.

In the end, the renovation ended up being way over budget, but it was worth it. When we started the job, the original value of the property was probably $1.5 million (this was seven or eight years ago) but I reckon that house would easily be worth $5 million today.

I really enjoyed that job. I was happy with the builder I was working for, I was happy with the clients – we all had a great relationship and it was a fantastic result. As the foreman, I was able to really get their vision in my head and I just made sure it happened the way they wanted.

WHAT TO LOOK FOR IN A GOOD CARPENTER/FOREMAN

What you want in your foreman will depend in part on what you want done, but there are a few guidelines that will most likely apply to everyone. I recommend you interview three or four people before you choose one and really take your time in making this decision. This allows you to consider several candidates before you settle on the one you want to hire.

Quality work and references

Your foreman will probably be your carpenter and should be able to show you two or three properties that he or she has build in the past. It's worth your time to take a close look at other builds he or she has completed.

If a candidate is hesitant to show you work done in the past, you should be wary of him or her.

In doing your groundwork, it's very important that you chat to a few previous clients of any potential foreman. You have the right to ask everyone you interview for references and why not?

Your project may cost you several hundred thousand dollars, so you should be very wary of any foreman who can't or is reluctant to put you in touch with his or her previous clients. These clients can tell you what they liked or didn't like about working with the person. Talking to previous clients will either put to rest any concerns you may have or give you an idea if the person is difficult to work with, aggressive or rough - but it's definitely worth your time to ask.

The right personality

It may sound a little hokey but when you're interviewing potential foremen, you should really listen to your gut. You need to be able to trust this person - and vice versa - so you'll need to form a good working relationship very early on. Only you will be able to know the type of personality that will work best with yours, and what appeals to someone else might not appeal to you.

Reliability

Your foreman will be essential in making sure your project stays on budget and on schedule, so pay attention to things like how well the candidate manages time and people before you hire the person. If someone is consistently late for meetings or doesn't seem to value your time before you begin the build, you may not be able to count on him or her to show up on time once the work kicks off and the clock starts ticking.

A willingness to listen

It's very important that your foreman listens to you and does everything to achieve the design you want. Of course, you are employing an expert in areas you're probably not too familiar with, but in the end, your foreman works for you and should cater to what you want. If you get the feeling that your foreman is not really listening to you, then think about finding someone else.

An aptitude for explanation

In working with your foreman, it's imperative that he or she listens to you, but it's equally important that you listen to your foreman and that you're willing to take direction and advice - it might be even more important for you to

be the good listener. You may have a particular design element in mind, but sometimes what we imagine just isn't feasible. Your foreman should be able to explain why and what options are available to you.

Also, if you want to help then let your foreman show you – watch and learn – and if you can't do it, know when to step away and leave it to the professionals. There will be lots you can do to help, but thinking that you can do everything is a dangerous mindset to have.

Insurance

In Chapter Four, we talked about all the insurance you should have prior to starting your build. Your foreman should be able to provide you with all of the required insurance as well, such as WorkCover, public liability and domestic building insurance.

The rules vary from one place to the next and will depend at least in part on the work you're doing, so check the requirements for your area and your project.

First aid

Throughout your build, you'll need to have someone on site at all times who has an adequate first aid certification.

If you find a foreman you like who doesn't have that certification, it would be worth your while to pay for your foreman to attend a one- or two-day course – and if you were qualified as well, so much the better.

An apprentice

It would be ideal for your foreman to have an apprentice or have access to other carpenters to help throughout the build. There will be times when your foreman can do the job alone but generally having at least one person to help is ideal. If that person can't be you, an apprentice is a great option.

Why? Apprentices are basically students, so you don't pay them as much as a fully qualified tradesman. Also, while two heads are sometimes better than one, you might rather your carpenter have an assistant or even two to help instead of having two equally-qualified tradesmen arguing over best practices.

Q&A

What if it's not working out with my foreman?

Having a good foreman is vital for any owner builder, especially ones who don't have much building experience. Your foreman will be the second most important person on the job site after you, but he's always replaceable. If at any stage you feel like things just aren't working out between you and your foreman, you can always fire him or her and find someone else. You simply have to pay for the work done for you so far and end the contract.

Changing foremen is a big thing, though, so you should think long and hard before taking this step. If you do sack somebody, you should plan to allow a week for the transition. It may cost a thousand or so dollars, but that's a smaller price to pay than to allow the wrong person to do the job.

Also keep in mind that if you're having a problem with your foreman, the problem might be with you – try to work things out if you can and remember – there's a right way and a wrong way to communicate. We'll talk more about getting the most out of your tradesmen in Chapter Seven.

WHERE TO LOOK FOR A CARPENTER/ FOREMAN

There are lots of places where you can find a carpenter/ foreman: newspapers, friends, online, on a local job site, a builders' association or advertise.

Most people get the local newspaper delivered to their house every week. Take a look in the classified section under carpenter or builder.

Ask your friends or co-workers for recommendations of people they've worked with before – but keep in mind you probably don't want to hire a friend or family member. If things go wrong with your build, you can damage a relationship and you may have real trouble rectifying mistakes.

Of course heaps of carpenters will advertise their services online. Do a search for carpenters or builders and see who's available to come to your area. Whether you're doing a big build or a small renovation, you can advertise your job, either in your local paper or online. Placing an ad on Gumtree (www.gumtree.com.au) or hipages (www.homeimprovementpages.com.au) isn't expensive and will put you in touch instantly with any number of available tradesmen. There are several builders associations out there, professional organisations designed to maintain high standards within the industry. Builders and tradesmen must hold the proper qualifications to obtain membership in a building association, so they're a good place to look for people to hire. Two of the best known are the Master

Builders Association and HIA, but they're not the only ones. Research what's available in your area.

Get in your car and you probably won't have to drive very far before you see someone out working. I found one of the best builders I've ever worked with on a job site. I simply stopped the car, walked up and said, 'G'day mate, I'm looking for a carpenter – are you interested?'

FIRST MEETING WITH YOUR FOREMAN

Before you hire your foreman – and before you make any payments – it's a good idea to have an initial meeting on site with your foreman so that you can go over everything that you require before you get started.

Discuss the role

From the very beginning, make it clear to your foreman that you are acting as the owner builder and that you want the carpenter on site to also help you run the project and oversee the build.

Discuss your plans

Your architect and engineer should have provided you with detailed plans. The architect's plans probably won't make a lot of sense to you or any other person who isn't a professional builder, but your foreman should be able to read a set of blueprints. If your foreman isn't completely certain about every element of the architectural and engineering designs, then consider getting someone else for the job.

Also, a good foreman should be able to anticipate certain obstacles simply by looking at your plans. Your architect and engineer should have pointed out and worked around any of your design elements that are structurally impossible or ill conceived, but your foreman may identify other potential difficulties. If there are any other documents or plans your foreman needs, make sure you provide whatever is required so that you can go ahead with the build.

Draw up a rough master schedule and budget

You won't be able to finalise this step until you've organised your tradesmen, but after looking at your plans, your foreman should be able to give you a rough estimate of how long your job should take and approximately how much you can expect to spend for each element of your build.

Your foreman can help you make those decisions about materials and design to help you get the results you want within the confines of the budget you have available to you. We'll talk more about organising tradesmen in Chapter Seven.

A Word to the Wise

Don't be difficult

For a builder, there's nothing worse than working for a bad client, but you should know that a bad owner builder can share a lot of qualities with a bad client.

About five or six years ago, I did a bathroom renovation for an older Russian lady who was very hard to work for. To begin with, she wouldn't let me park my car in front of her house – I never knew why. It was really annoying because as a tradesman, I'm constantly going back and forth to my vehicle for tools or other supplies, but she would make me park three doors down.

Also, every time I came into her house, she'd have me take my shoes off, regardless of what state they were in. I know she didn't want me to leave her house a mess, and I guess other builders might, but it would have made more sense to me if she had let me clean up after myself at the end of each day.

But the worst was that this lady would just sit there and watch me all day. She didn't have a clue what I was doing, but she'd talk to me all day. She would try and get information out of me about the company I was working for – she would bag the company, trying to get me to say something bad about them. I probably wasted two or three hours a day talking to her, which held me up a lot. All I wanted to do was to do my bloody work.

She was so bad that I dreaded going to her house and I didn't want to be there when I got there. I take pride in my work so I did the job she contracted me to do, but she certainly didn't get the best out of me. Over the course of two weeks, I reckon she would have wasted over twenty hours of my time, which adds up.

As an owner builder, you want to be involved, but it will work out better for you if you're not difficult to work with or to work for.

PAYING YOUR FOREMAN

Another thing you'll want to discuss in your initial meeting with your foreman is what your foreman expects of you - namely, how you plan to pay - and that will probably depend on your project. There are a few options that make sense.

By the hour

If your build will involve less carpentry work or you plan to be available to oversee more of the project yourself, it might make sense to pay your foreman an hourly rate. This method is best if you anticipate that your foreman won't be working at your job site every day, or if you want to break up the time spent doing carpentry work and the time spent acting as a foreman.

By the job

If you're working on a big build and you anticipate your foreman doing a lot of work every day, it might be better to agree on a fee for the entire project. In this case, you're basically hiring the foreman to work for you for the entire period of time that your build is happening. This method works best if your foreman is willing to kill two birds with one stone, doing carpentry work and acting as a foreman at the same time – which is basically what you're looking for.

By the week

If you want your foreman to be available to you for the duration of your project but you're not sure how long it's going to take, you may want to agree on a weekly salary that you can pay from the time the work begins until the build is finished – however long that takes.

However you decide to pay your foreman, you should draw up a contract in writing so that both of you are protected legally in the event of a dispute. Also, if your foreman has an apprentice, now is the time to decide how you're going to pay the apprentice as well and draw up contracts accordingly.

CHAPTER SUMMARY

Your foreman is one of the most important people you'll hire for your project, so choose carefully. A carpenter is usually involved in just about every step in building a house, so it makes sense to hire a carpenter who can also act as your foreman at the same time.

You can hire a builder to act as your foreman, but if you do, make sure he or she understands that YOU are the owner builder. You cannot simply leave everything in his or her hands because you are ultimately responsible.

In addition to doing all of the carpentry work (or finding someone who can), your carpenter/foreman will also help you find and oversee your tradesmen. If you want, you can have your foreman liaise between you and the contractors, land surveyor, architect and/or engineer plus your foreman can help you draw up a budget and master schedule.

Your foreman doesn't *have* to be a carpenter, but you'll probably have a carpenter on site throughout the build anyway, so you may want to combine the two roles.

You can find a carpenter/foreman (or a builder to act as foreman) in a number of places. Check out your local paper, search online, ask your friends or advertise your job. You can also check the relevant builders' associations or drive around and approach someone you see working.

A good carpenter/foreman should be able to produce quality work, provide references and have the required insurance and first aid qualifications. You want someone who is reliable and it helps if an apprentice is part of the deal. Finally he or she should demonstrate a willingness to listen and an aptitude for explanation, and have a personality that suits yours.

If at any stage you feel strongly that things aren't working out with you and your carpenter/foreman, you can always replace him or her. It can be a difficult change to make, though, so it's better if you can pick the right person the first time.

In your first meeting with your carpenter/foreman, you'll need to discuss every role you'd like for him or her to do, discuss your plans and draw up a rough schedule and budget.

Prior to commencing work on your project, you and your carpenter/foreman need to agree on the payment structure – usually either by the hour, by the job or by the week – and draw up a contract accordingly.

WHAT COMES NEXT

You've got pre-approval for your financing, you have your licences and insurance, you have plans and engineering and you've found a foreman. You should have been able to draw up a rough budget and time frame, but now you'll want to hire the contractors and formalise more specifically how you plan to spend your money and your time.

CHAPTER SEVEN

ORGANISING YOUR TRADESMEN AND YOUR INSPECTOR

You're getting close to the time when you'll start building, but you're not there yet. Before any work can happen, you need to know which tradesmen to hire, when to have each of them come and how much you'll need to spend on materials and labour for all of them.

Why should you go with professional tradesmen? A lot of people think they can do plumbing, painting and electrical work themselves. They look at the finished product and think, 'How hard could it possibly be?' The answer is usually, 'Harder than you think'.

You may be tempted to look at the tradesmen's hard hats and work boots and think you're smart enough to do those jobs yourself, but think again. Those tradesmen will have had years of training, including course work and apprenticeships – most of which require at least four years to complete.

You wouldn't expect them to be able to come into your place of business and do your job without any training or experience, so don't assume you can do theirs.

TRADESMEN YOU'LL NEED TO HIRE

Let's take a look at a few of the tradesmen you'll probably need to hire to build your house and the qualifications they should hold. (Notice I said 'a few' – depending on what you want to do with your project, you may need to hire other tradesmen in addition to the ones I've included below. Your architect and/or foreman will be able to advise you.)

Concreter

As the name implies, your concreter deals in pouring and spreading concrete, plus also compacting, finishing and curing the concrete. If you need to hire a concreter for your build – for pouring a foundation, a driveway or a swimming pool, for example – you'll want one with the relevant qualifications. The specifics are different for each state, but to become licensed, most concreters will have taken a course to achieve a Level III Certificate in Concreting. They will have also undergone a minimum of four years as an apprentice.

Carpenter

In Chapter Six, we talked about how your carpenter will probably be involved in nearly every step of your project, which is why he or she is a good candidate to act as foreman.

In general, a carpenter builds or repairs all the bits of the house made from wood such as frames, staircases, door frames, partitions and rafters, plus other elements including cabinets, vanities, cladding and cement sheeting. They can also do other jobs like inspecting your foundation and fitting out the last finishing touches of your house.

There are varying degrees of certification for carpenters depending on what they're trained to do, but in general, you want to look for a carpenter with Certificate III in Carpentry. Any carpenter with this certification has probably completed four years of apprenticeship (although it's possible to finish an apprenticeship in less time – I did).

Plumber

When you think of a plumber, you probably think of toilets and sinks. In fact, plumbers deal with a lot of other services too, including gas, water, sewerage, heating systems and even roofs.

To become a licensed plumber, a tradesman will need to obtain a registration or a plumber license, which involves earning a Level III or Level IV Certificate in Plumbing.

These levels of certification require at least two years of training and plus at least four years as an apprentice.

Electrician

Electricians, who (obviously) deal with all of the electrical components of your build, are always in high demand, mainly because their work is so dangerous. Your electrician should hold a Level III or Level IV Certificate and would have had to pass quite a lot of exams to become qualified - and of course these qualifications are usually in addition to a lengthy apprenticeship.

Working with electricity is a highly-specialised area of expertise, and as such it is closely regulated so check for what's required of an electrician in your state.

A Word to the Wise

Don't get a shock

Electrocution is a common accident in the building industry and it's never nice. I've probably had four or five shocks myself over the years. One time, I was walking past an electrician who was fitting off a fuse box when it sparked and blew up in his face. It took about six months for him to regain his eyesight and he had to go on medication to get his heart beat back to a regular rate. Accidents like that one can happen to the most skilled professional electrician - which means they are worth every cent and you should *never* try it yourself unless you're fully qualified.

Plasterer

It's easy to confuse the job of a plasterer with that of a painter, but make no mistake – they're two totally different things. A plasterer will apply gyprock and cement sheeting plus finishing coats to create a smooth look to walls, both inside and out. The plaster protects the framework and creates a flat surface perfect for painting.

This is one of those detail jobs that simply *must* be done correctly, so you'll want to find a plasterer with at least a Level III Certificate in Plastering, which will include the same four years of apprenticeship as most other trades.

Painter

Painting is one of those jobs that a lot of people think they can do. On *The Block*, in fact, everyone does painting. But those contestants have one thing you don't – me standing over their shoulders, making sure they get it right.

In most states, a qualified painter will need to hold a license, and getting that license requires a Level III Certification in Painting plus the four years of apprenticeship and passing scores on all the requisite exams.

Still think you should try doing those professional jobs yourself?

I have one thing to say – please don't. Unless you have some real training or skill, you should leave those jobs to the professionals. There are other jobs not listed here like landscaping and tiling that you'll probably also need to have done, but you get the idea. You probably don't have the qualifications to do this kind of work.

Having said that, there are some jobs you can (and should) do if you have the time, a hard hat and protective clothing. We'll talk more about what you can do to help in Chapters Eight, Nine and Ten.

WHERE TO LOOK FOR YOUR TRADESMEN

You'll look for each of your tradesmen in the same places you looked for your carpenter/foreman, namely the paper, by word of mouth, online or out working.

You can also advertise and ask your foreman for recommendations. If you can, it's a good idea to work with someone your foreman trusts.

If your foreman isn't able to suggest some tradesmen, each trade has its own professional organisation - not all tradesmen are members of their professional organisation, but it's a good place to start.

WHAT TO LOOK FOR IN A TRADESMAN

While each tradesman should have the qualifications, licenses and certifications I mentioned above, there are certain other traits you should look for in every tradesman - including your carpenter/foreman.

Reliability

It is paramount your project stays on schedule, so if a tradesman doesn't respect your time and consistently shows up late, your build could start dragging on and costing you money.

Quality work

As with your carpenter/foreman, your tradesmen should be able to show you examples of their work and provide you with references.

ABN

An independent tradesman is likely to run his or her own business. If so, the business should have an ABN, or an Australian Business Number for tax purposes. A good rule of thumb: no ABN, no job.

Insurance

The requirements vary from state to state, but in general each tradesman should have public liability insurance, WorkCover insurance, domestic building insurance and/or a contractors insurance certificate of currency. Check out the rules for your area.

White Card

In Chapter Four, I said that everyone who does construction work in Australia is required to have a White Card, which indicates the completion of General Safety Induction Training. Each of your tradesmen should be able to produce a valid White Card.

Warranty

All tradesmen should be able to provide you with some way to guarantee the quality of their workmanship in the event that something goes wrong in the future. Think about including some sort of warranty clause in your contracts.

Safe Work Method Statement

This essential document basically outlines how each tradesman plans to do his or her job safely. This statement is mandatory so don't let anyone begin work on your property without it. If you do and the Work Safe representatives show up to do an inspection, you'll find yourself in serious trouble.

HOW TO GET THE MOST OUT OF YOUR TRADESMEN

Back in Chapter One, I said that anyone who can't control his or her temper is not likely to make a very good owner builder. Even if you have your foreman liaise between you and your contractors, there are a few guidelines you can follow to maximise the relationship you have with everyone you hire to work for you.

Be polite

If you don't like something, there's a way of saying it. You can choose to get really grumpy and create drama, or you can nicely say to someone (maybe even with a smile on your face), 'I just don't think that's right – talk me through it.'

You don't have to rant and rave. If you do, you might make the person work harder but more than likely you might just make the person resent you. And the last thing you want are people working on your site who are not very happy with you: you're likely to end up with a shoddy job.

Be specific

You'll be forgiven for being passionate about the design you've come up with for your house, but it's imperative you talk to your tradesmen in specific terms. If you're vague or uncertain, they won't be able to get your vision out of your head and make it a reality.

Be flexible

At the same time, the things you've imagined aren't always possible, practical or within your budget. Most of the time, your tradesmen should be able to work to your specifications, but you need to be prepared to listen to the advice of the experts.

Be professional

I've said it before, but I'll say it again. Try to not use friends or relatives. Business is business. Unless you have a friend you can totally trust who might do you a deal, it's not a good idea – you may be able to fire a tradesman but things can turn ugly if that tradesman comes to Sunday dinner every week. And even if you have friends or people who might do you favours, they might want something back in return, so really be careful.

Be firm

As the owner builder, you're always the one in charge. You need to be able to say what you want and not let other people run over you but at the same time you don't want to cross the line into being confrontational – they like to call me 'Big Bad Keith', but this is the line I try to walk on *The Block*. If you don't think a tradesman is working out, you can always determine what percentage of work has been done in relation to the contract, pay that person accordingly and send him or her on their way

Be organised

Before you engage any tradesmen, always get written quotes and contracts. You'll probably need these quotes for your financing anyway, but even if you don't, it will only work to your advantage if you have your quotes in writing. You also need to ask for written receipts for all the work you have done on your house, especially if you're doing renovations. This documentation will help you to prove the value of your home and will also cover you if there's ever a dispute. Just keep an owner builder file or folder and make sure everything goes in it.

A Word to the Wise

Don't even try to know it all

The first time I took on the role of foreman, I was only 26 years old. A lot of the guys on the crew were older than I was so it was a little difficult for me to be in charge. If you're feeling a little unsure about managing a crew of tradesmen, keep in mind that you have to stick to your guns, but you also have to listen because a lot of those guys are going to be more qualified than you are.

If there's something you don't know, don't pretend you do. Throughout my career – from my first foreman job until now – I have relied on my network quite often because as a tradesman, I can't be an expert at everything. No one can. I'd say I'm an expert in my field of carpentry but with plumbing or electrical questions, I always rely on my experts. If I ever find myself in a circumstance and

I don't know what to do myself, I'll ring up a professional to come in and sort things out. You should do the same.

DON'T JUDGE A BOOK BY ITS COVER

In hiring tradesmen, you really can't judge people on appearance. Sometimes a person might look fantastic and sound great on paper, but then when they show up in their crisp clean uniform, their work isn't what it should be. On the other hand, another person might look rough as guts but produce the best work you've ever seen.

Also, keep in mind that tradesmen are human. There have been plenty of times when I've used fantastic tradesmen who have been going through a divorce or a death in the family. They'll have a week or two when their mind isn't on their job. In other words, if a tradesman has an off day, that doesn't necessarily mean he's not a professional.

When you start working with someone, you'll realise they have both strengths and weaknesses. I once knew an electrician who was great at his trade, but he would constantly forget things, like whether or not he'd been paid. Sometimes he would be unreliable and not turn up for a few days, but when he did show up, he did a really fantastic job. For me (and the time frame I was working around), it was worth keeping him on my crew, in spite of his shortcomings. As an owner builder, you may have to manage tradesmen with certain faults and it will be your call to decide how you want to deal with them.

A Word to the Wise

Be prepared to overlook a few flaws

I used to work with a tiler who had the worst body odour I'd ever smelled. He absolutely reeked. He smelled so bad that I didn't want to be in the same room with him, but I had no choice. I would have to work sometimes for hours in a bathroom with the guy. At the same time, he was a lovely bloke and a great tradesman so rather than kick him off the team, I just put up with his smell. You'll find there are many people who appear rough and burly but they're actually good tradesmen and quite clever, so don't dismiss them too soon.

DRAWING UP A MASTER SCHEDULE

You've heard the expression 'time is money' and that's most definitely true when it comes to building. Get your schedule wrong and you may end up paying for materials that won't get used in a timely manner or you'll be paying for tradesmen to stand around doing nothing rather than working. To keep your project on track, you should make a master schedule that covers the entire build from 'go' to 'whoa'.

Use a whiteboard

Sit down with your foreman and work through every job that will need to be done for your project to reach completion then put those jobs in order according to when they need to happen. Your foreman should be able to advise you which jobs will need doing, in what order and about how long each should take. If digital is more your style, draw up a soft copy document in something like Excel – just be prepared to make a lot of changes.

Pencil in your contractors

Start by choosing each tradesman with care. Your foreman should be able to suggest several, but you can do your own research to find the best person for each job. Everything will need to happen in a fairly specific order, so go through your list once and make tentative bookings for everything, including the carpentry your foreman will do. Each tradesman should be able to give you a rough idea of how long the job will take – then you can allow a margin of an extra day or two for each job.

I say tentative because, at this stage, certain dates may need to change as you speak to each contractor. Also, you may need to make changes based on what you can and can't afford. As always, try to get at least three quotes for every job.

Find materials

A good way to get a baseline idea of the cost of materials is to send off your plans to the major hardware stores. They can provide you with a quote for all the building materials for your whole house, including timber, plaster and fittings. They can also tell you when these materials can be delivered. You may be able to find a better price or delivery date if you look around, but the major hardware stores are a good place to start. Include these delivery dates on your schedule.

Confirm your contractors

Once you've got a tentative range of dates for each job that needs to be done, you can go back and confirm each booking with each tradesman. Make sure that every contractor is happy with your master schedule before you get started. If anyone isn't happy with the amount of time allotted to him or her, amend your schedule accordingly. You will have to be a little flexible when the build actually starts – some jobs may get moved forward or back by a day or two for unavoidable reasons – but for the most part, you'll need to make sure everything goes according to your plan.

Making a master schedule may sound a little scary, but your foreman should be able to help you. It may take awhile to draw up a feasible schedule, but it's worth the time to get that schedule right and make sure each tradesman feels comfortable with the amount of time allotted for the work to be done. If a company doesn't have enough tradesmen to get your job done in the amount of time you need it done, you may need to look for a different company.

If you don't feel comfortable making the big decisions and you feel like your foreman is capable, then you may want to trust him or her to make some of the hard choices for you but you should always be aware of everything that's happening in relation to your project. If you're not, your foreman will become the builder instead of you.

DRAWING UP A MASTER BUDGET

As you're in the process of drawing up your master schedule, you'll be getting quotes for labour and materials at the same time. For this reason, it makes the most sense to draw up your budget alongside your schedule. I could show you a sample budget here, but it would likely be unrealistic and possibly confusing. Instead, here are a few guidelines for drawing up your own.

Look at your entire master schedule and work out when you'll need to pay for each individual job. In general, tradesmen expect to be paid within a few days of finishing their contracts, sometimes as quickly as seven days, sometimes as long as 30 days. This expectation varies widely, however, so you can (and probably should) include when you'll pay your tradesmen in the contracts you have with each one.

You can draw up a second grid that matches your master schedule, but a better method would be to simply include the amounts you'll need to pay as a separate line or in a different colour on each day of your master schedule.

Often your tradesmen will include the cost of materials in with their quotes. If you're buying materials elsewhere, include those costs and delivery dates along with payment dates on your schedule according to when they should come due. Once you know exactly how much you plan to spend, on what and when, you can go back to your lender and formalise your loan. Using your master schedule with the payment figures included, you can get a very good idea of when to schedule your progress draws on your loan. Arrange these payments accordingly with your lender.

Which tradesmen you'll hire and which materials you'll use will most likely come down to your budget. Your financial plan will become more certain when you order your materials, which will probably happen at the same time as when you make finally book and lock in times for your tradesmen.

This is the stage at which you'll need to make some final decisions about which materials and techniques you want to use, but if there are a few elements you're still not 100 per cent sure about, don't panic. The beauty of being an owner builder is that you can still make changes even after the build has started without too much drama. The more accurate your proposed budget is, though, the better.

Where should I plan to spend money? Where can I save?

As you book your tradesmen and order materials, you'll discover that you can either splurge or save at nearly every step along the way. Some rooms of your house, however, are worth the money.

Kitchens, living rooms, bathrooms (especially the en suite) and master bedrooms are usually the selling points of any property. These are the rooms where most people spend the majority of their time so don't scrimp on these areas.

If you do your homework, however, you can sometimes shop around and save money on materials. Sometimes contractors can get you a better price on materials because they can buy in bulk, but you can often get things like tiles, floorboards, kitchen and bathroom taps and other fittings for less money by shopping around. You might also consider using a cheaper framing or cladding material like foam and render instead of double brick. If you're over budget, talk to your foreman or architect about where you can save.

ORGANISING YOUR INSPECTOR

You're getting close to 'go' time, but there's one more essential step you need to take before you start setting up your build site – organising your building inspector.

Finding and organising your tradesmen is likely to be a lengthy process, one that won't happen over night. At the same time that you're booking your contractors, you can also find the building inspector you want to use. In some states, you can engage these professionals through your local council, or you may be able to hire them privately.

Inspector or surveyor?

A building inspector is sometimes called a building surveyor. Just so that we don't get confused, for the purpose of this book I've made reference to a land surveyor and a building inspector. If you hear the term building surveyor used elsewhere, it's still the same role as the building inspector.

A Word to the Wise

Get it checked out

About fifteen years ago, I was working as a carpenter for a friend who was a builder. We were building a house and we went all the way through the framing stage and on to the **fitting off** and finishing stages. As will happen with every newly-built house, my builder friend had the inspector come to perform the final inspection and get the **Certificate of Occupancy**. It was at this stage that he found out his frame inspection hadn't been ticked.

Fitting off (or fit-off stage) is the final detail work that makes a house liveable. It is the last stage of the building project and involves the last of the carpentry, electrical work, plumbing work, painting and flooring.

The **Certificate of Occupancy** is the document that proves your house has passed all stages of inspection – without it, you won't be able to insure your home.

We thought the house was finished and we had expected to pass the last check, but when the inspector turned up, it was a different guy to the one we'd been working with throughout the project. As it turned out, my friend had had a relationship with the previous inspector, but the first inspector had left the company and he hadn't documented everything properly prior to leaving.

It should have been a small matter of paperwork for the previous inspector, but leaving that documentation undone ended up having disastrous consequences for my friend. He thought the house was finished and the client would be able to move in within a few days. Instead, he heard the new inspector say, 'Sorry, mate, but there are a few spots on the frame I still need to check.'

For the sake of a minor bit of frame inspection, we had to strip the house back to the frame. All total, it cost about $50,000 dollars to rectify that mistake – probably $27,000 for replastering and another $13,000 for painting. I think the only things we were able to save were the ceilings.

Because my friend was the builder, he had to wear the cost of all the work we had to do again, and fair enough – it was his fault, even though he was an experienced professional. He managed to survive that mistake but it's the sort of oversight that can put a builder out of business. If you do the same as an owner builder, the cost of repair will fall on you.

The role of the inspector

The building inspector will be one of the most important people to come onto your jobsite. Contestants on *The Block* are often terrified when they see the inspector coming because they aren't able to proceed with their project until they get 'the stamp of approval'. The same will be true for you.

Think you can save some money by skipping inspections? Think again. Without the proper inspections, you can't get your Certificate of Occupancy and without that, you can't get house insurance. It may seem like a bit of a hassle, but building inspectors are there to protect you. They make sure that your property is being built to code – a step that's especially important for amateur owner builders who may not know what to look for as work on their property proceeds. Those inspections will tell you that your carpenter and other tradesmen are doing the work they should be doing.

When hiring an inspector, look around. Prices and procedures vary from one person to the next. Where you live may also be a factor. As with hiring any other tradesman, do your research and get the opinions of people who have been there before contracting your building inspector.

A Word to the Wise

Embrace your inspector

At the end of the day, the inspector is there to help you. Let's say you build your house, shed or fence too close to your boundary line. You may only be a few centimetres off the mark, but if you're caught, you may have to pull down your entire structure and rebuild – and it won't be the inspector who enforces that requirement, it will be your local council.

Think you can get around those rules? Think again. The council has these laws for very good reasons. It might not make sense to you but it's all about how your building will affect your neighbours and how it will impact the entire neighbourhood visually. There's no point in trying to fight against those rules – you're not going to win.

Even today, I don't always understand the rules that regulate the way we have to build, but the people on your local council have been looking at every aspect of building and taking it into consideration for a long time, so don't just think you're clever or smarter than everyone and that you can get around it. Instead of trying to avoid them, you should welcome your inspector's remarks.

When to inspect

There's not one simple answer as to when your inspector will want to look over your build. The number of inspections you'll need will depend on what you're building and how. The same is true for what stages in the build the inspector will need to come. In general, you can expect to have an inspection of at least four critical stages of your build – the foundation, the frame, the lock up and the final check.

The same is true of renovations. A small renovation will probably incur fewer inspections, but they're still necessary. Check with your inspector prior to your build – find out exactly how often and when inspections need to be done to assess the work on your property.

Each inspection is determined before you begin the project. Your inspector will look at the job, look at your plans and tell you how many inspections you'll need. You'll usually organise those inspections and pay for them up front. If you need more, you'll pay for them as you go, usually paying a couple of hundred dollars for each.

If you fail an inspection, your inspector will give you a list of items that need to be rectified – he'll then have to return to reinspect and this second inspection may cost you a second fee. One thing is certain: every inspection you pass will get you another step closer to that all-important Certificate of Occupancy. Without that final verification, you can't move into your house.

The unhappy inspector

Problems are sure to arise in any building project, but if your inspector is constantly unhappy with the work that's being done on your property, you should really wonder why.

The first place to begin to look for a solution is with your foreman – you've hired this important person because he or she has more expertise than you. It's up to your foreman to make sure the tradesmen are doing quality work. If the inspector sees that the job isn't being done properly by any of your contractors, the issues should be brought to your attention as soon as they happen so that they can be fixed.

If a conversation isn't enough to remedy the situation, you may need to release any substandard tradesmen from their contracts and find replacements. You don't want to have to pay to have the same job done twice, so it's a good idea to include a clause in every contract with every tradesman that the work must be done to code.

You may also need to consider the quality of your foreman's work. Part of the job is to supervise the work of the other contractors. If you aren't able to spend much time at the job site, your foreman might take advantage of your absence and cut corners or give your build less attention than it needs.

WHAT TO DO IF IT ALL GOES WRONG

If you're not sure about any aspect of the build, ask! It's better to confront the situation and get it right before the inspector arrives. Otherwise, your project may not meet code and fail the inspection. If that happens, you'll have to tear down the inferior work and start over – which is almost sure to cut into your budget and throw off your schedule.

If your inspector consistently finds flaws, you may find yourself in real trouble. As much as you can, you should keep an eye on the project yourself, but if things are starting to slip out of control, it may be worth your time and money to hire an independent builder for a few days or weeks and pay him or her by the hour to help you get back on track.

Your inspector isn't there just to make sure you abide by the council's rules. He's there to make sure your house is being built properly. Leave him or her out of the equation and you may be very sorry.

A Word to the Wise

Beware unlicensed tradesmen

Just a few years ago, I had my own carpentry business and a builder rang me, looking to hire on a few of my guys to help him with an extension he was building for a client. It was a pretty straightforward project, just a 12m square extension – easy.

I went down to the suburb in southeast Melbourne and started running my eye over the job. At this stage, the frame was finished and the house was bricked up and ready for plaster, fit off and flooring. As soon as I arrived, however, I could see that the stumps were too far apart. Normally, stumps should be 1.5m to 1.8m apart, depending on your soil – these stumps were 2.4m apart, probably 30 or 40 per cent off where they should have been. When your stumps are too far apart, your entire house will be much less stable than it should be.

I also noticed that the builder in question had used a soft wood instead of a hard wood for the **bearers** and floor **joists,** and that the cavity between the framing and the brickwork was too big. They had been on the verge of fitting off but these were some very serious structural problems, and the entire extension had to come down.

The guy said he was a builder but to this day, I don't know that he was ever really qualified. He had made an agreement with the clients – he told them he didn't have his license and that it would be better if they didn't get building inspectors involved. He also told his clients that he could do the job 'behind the scenes' and no one would know about it. He insisted they didn't need to worry about it.

When I discovered all the problems, I had to tell the clients. It probably cost over $70,000 to repair all the flaws. The expense should have fallen to the builder, but the builder and the client were in a funny situation because they had a verbal agreement but they didn't have the proper contracts in place. I think the clients and the builder ended up splitting the cost but it was very tricky.

Both the builder and the clients wanted me to stay involved in the project so against my better judgement I did. I fixed up the house – for three weeks, I had seven or eight guys working seven days a week until seven o'clock every night. We pulled down the brickwork, we pulled down the framework and the floor and we rebuilt the whole thing.

Other than my guys, I don't think anyone who worked on the project was qualified – a situation that never would have happened if the clients had been using an inspector. He would have seen that the stumps were wrong before the concrete was even poured. That massive do-over could have been fixed for a fraction of the cost if the builder was legit and inspectors were involved.

In the end, the builder in question never paid me. My component of the work must have been about $23,000 or thereabouts. I paid the guys working for me to fix it up but I just wore the cost. In other words, it really pays to play by the rules.

Bearers are part of the subfloor – these lengths of hardwood sit directly on top of your stumps to create a solid base for your house.

Joists are lengths of hardwood that lay perpendicular on top of the bearers directly beneath them. The joists are closer together than the bearers and create a sturdy grid on top of which you'll lay your floor covering.

CHAPTER SUMMARY

- Even though you're the owner builder, you're probably not qualified to do most of the specialist building work yourself.

- You'll need to hire a number of tradesmen to build your house including (but not limited to) a concreter, a carpenter, a plumber, an electrician and a plasterer. Each of these tradesmen will have completed several years of coursework and apprenticeships to obtain their licenses.

- You can find your own tradesmen by looking in all the usual places like the newspaper, online or with the professional organisations, but your foreman should be able to make recommendations.

- A good tradesman should demonstrate reliability and quality work. Additionally, all tradesmen should have an ABN, all the proper insurance and a White Card. They should also be able to guarantee their workmanship with some kind of warranty and show you a Safe Work Method Statement.

- You can get the most out of your relationship with your tradesmen and your foreman if you can be polite, specific, flexible, professional, firm and organised.

- It's best not to judge tradesmen on appearance alone - some will look very neat, clean and sharp but do substandard work while others may come across as sloppy or gruff but do a great job. Getting the right person to do the best job is what you want: not Mr or Ms Neat.

At this stage, you should be able to sit down with your foreman and a white board and draw up a master schedule. You'll create this schedule by organising the days that each of your tradesmen will come and how much time each will need to complete their part of the project.

At the same time you create your master schedule, you should also draw up your budget by getting quotes from all of your tradesmen for labour and materials. Then you can add these figures to your master schedule so that you'll know the best time to schedule the progress draws of your loan. Once you have all of your quotes in hand, you can go back to your lender and finalise your financing.

It's worth putting more money into your kitchen, lounge room, master bedroom and en suite as these rooms are usually the selling points of a property. You may be able to save some money, however, on certain materials and fittings.

Prior to starting your project, you need to organise a building inspector. The inspector is sometimes called a building surveyor, but for the purpose of this book, I'll use the terms, 'inspector' or 'building inspector' so we don't confuse this role with that of the land surveyor.

Your inspector is vital to your build because this is the person who insures your tradesmen are producing quality work that meets code.

When and how often your inspector will come depends on what you're building – even for renovations – but in general, the inspector will check the foundation, the frame and the lock-up before performing the final check and giving you a Certificate of Occupancy.

If your inspector is consistently unhappy, something is wrong. Any flaws an inspector finds will need to be fixed, which will cost you extra time and money. You may need to consider releasing certain tradesmen from their contracts and replacing them. If all the tradesmen are falling short, you may also need to reconsider your foreman.

If things go really wrong, you may want to hire an independent builder for a short amount of time to help you get things back on track.

WHAT COMES NEXT

You've made a lot of decisions by this stage. All of your plans should be in order, from your design and finance to your tradesmen, your materials and your documentation. Your schedule and budget are in order and everything is ready to go. It's finally time to prepare your job site and get building.

CHAPTER EIGHT

THE FOUNDATION STAGE

By this point, you will have already put in a lot of time and energy getting everything ready for your build. You've done your planning, you have a budget and a timeline in place and all of your contracts are in order – it's time to get started! The good news is that, if you've done everything properly up to now, the actual build should go relatively quickly.

GETTING THE JOB SITE READY

The first thing you need to do is organise your job site. You'll have a lot of people coming in and out of your property, probably for several weeks, so you need to make sure everything is in place before they arrive so that they can do their jobs to the best of their abilities.

Looking after your people on site

It's vital that you provide a safe work place for everyone who comes on your job site.

First aid

On the job site, the foreman is in charge of health and safety. If your foreman isn't already certified, you should pay for your foreman to get a first-aid certificate. There will be different requirements depending on where you live, but usually a Level II or Level III Certificate is the requirement. If you're not sure, you should check because if there isn't a qualified person on site, you as the owner builder are the one who's liable. You should also have a good first-aid kit on site.

Public liability

As an owner builder, you need to arrange public liability insurance, which protects you if anything happens to the public as a result of your build. If you didn't organise this insurance during your documentation stage, you should do it now.

Toilets

You definitely want your tradesmen to have access to a proper toilet. You can hire a portable one but I think the best thing is to have a plumber set up a make-do toilet and run temporary pipes back to your sewer.

Portable toilets can be quite smelly and a lot of people don't like using them, including myself, not to mention they're expensive and if you use them long enough, they'll fill up. A temporary toilet is going to see you through your job, save you money and give you a better result.

Other amenities

Your tradesmen will be spending a lot of time on your property, so you want to make sure they're comfortable. Back in the old days, nothing was usually provided for workers but these days we treat people a little better.

In addition to your site toilet, you also want to set up an area on your job that's under cover, a place for your tradesmen to have morning tea and lunch. It's up to you to decide how far you want to go, but it never hurts to look after the people who are building your house.

Site office

If you're doing a big project, you might want to set up a little makeshift site office on your property. A site office is especially useful if you're starting with an empty block of land. Also, if you're living in your house during a renovation, you may prefer to provide a site office where your foreman and tradesmen can gather just for the sake of your privacy. If you do decide to use a site office, make sure it's not going to be in the way of your build. Think about where you're putting your site office in relation to your excavation and your build, and make sure you don't have any pipes or power lines running underneath it, so you don't have to move it later.

Access

Does your house or property have easy access? If it hasn't, it will incur extra cost, normally an extra ten per cent, because you'll need to crane materials in and out.

You'll also need to have a clear access point for an ambulance. If someone gets injured, rather than having the paramedics coming in and climbing over the fences and such, you want to have good access to the front or the rear of your job just in case of emergency.

Temporary water

You definitely need to have running water on your job. In addition to the toilet, there are heaps of jobs that will require water such as mixing concrete and mixing tile glue. You'll also need water to test plumbing pipes and you'll always need water for cleaning up.

Temporary power

The easiest way to provide the power your tradesmen will need to complete the build is by setting up a **builder's pole**, which is separate from your existing power and supports a big power box to run all of your leads out of.

Used only for tools or a site office, every power pole has a safety switch so it reduces the chance of electrocution. It's probably the easiest, cheapest and safest option – if you're using a power tool and you drop it into a puddle, rather than giving you an electric shock, just the power is killed.

An alternative to a temporary pole is a temporary generator. A 5000-watt gas generator should be big enough to handle a fairly large project. It's not ideal but if you don't have any other option, the generator is a good way to go.

Fixed ladders and handrails

If you're building up another floor, you need to have ladders securely fixed into position, and all of your ladders should extend a meter over the floor or the surface that you're working on. Also, any time work is happening on your roof, you'll need to install handrails to reduce the risk of anyone falling.

Security fences

Before you begin building, you need to secure your entire property by enclosing it in temporary fences because you really don't want to have people prowling around your build who don't belong there. Say you've got a hole in the ground on your property a metre or two deep – now let's say someone breaks into your property overnight. If they fall down that hole and break their neck, you are liable for their injury.

Safety first

Any time you (or others) go on the build site, be sure to wear gloves, safety boots and other protective clothing. Also, listen to the experts and utilise **PPE** – personal protective equipment – such as safety glasses, ear protection and a hard hat.

A **builder's pole** is a temporary power pole that is separate from your existing power, supports a big power box to run all of your leads out of and has a safety switch to reduce the risk of electrocution.

PPE stands for personal protective equipment. It refers to clothing designed to keep you safe such as hard hats, gloves, safety glasses, safety boots, ear protection and high visibility vests.

A Word to the Wise

Don't fall for it

It is vital you take safety seriously on your job site. When I was in my mid twenties, I saw a guy fall about ten metres. He'd been up on a ladder and he hadn't tied it properly. Also, the ladder was positioned on a tile floor and it didn't have any rubber feet.

When the ladder slipped out from underneath him, the guy had nowhere to go but straight down onto a pile of big metal air conditioning ducts. He was in his mid forties, I think, and an experienced guy, but he broke his leg and suffered massive cuts. In other words, even qualified tradesmen can have accidents, so make sure everyone on your job site is being careful – including yourself.

Getting rid of rubbish

Whether you're doing a renovation or you're building a house from scratch, there will always be a large amount of rubbish, so you'll need to have bins on site early on. All of your building rubbish can go into your skip including construction and demolition waste, cardboard, paper, metal, wood, dirt, tiles and glass.

Certain things can't go in your skip such as pressurised containers, electrical appliances, food or kitchen waste, human or animal remains or waste, asbestos, gas, batteries, oil and hazardous materials like chemicals, petrol or paint. If you have this type of waste, you'll need to contact your local council or research online and find out the best way to dispose of it properly in your area.

You should receive a comprehensive list of everything that can and can't go in the bin when you hire it, so make sure you check it carefully. Also, make sure your bin is on your own private property – if you need to leave it on the street, you may need to obtain a permit or face paying a fine.

Depending on what you're doing with your project, you may need to have your skip emptied several times throughout the course of your build. Be prepared to have the skip removed as soon as it is full and returned to you empty as quickly as possible.

Planning for your utilities

Now is the time to sort out your gas, water, electricity, sewerage, heating/cooling, telephone/Internet and any other utilities you anticipate needing once your house is finished.

Renovation

If you're doing a renovation and you need to demolish all or part of the standing structure, make sure every single one of your utilities gets turned off. The gas is perhaps the most important. The pipes will be in place but there should be no gas running. If an excavator comes in and rips up a live gas pipe, you've got a disaster on your hands. You could kill people, you could have a whole street evacuated – trust me, you don't want gas leaks. With electricity, you don't want any existing power in the ground. Your electrician might be able to isolate certain areas, but you're probably better off shutting down all the power. You just don't want any accidents.

Undeveloped land

If you're building on a plot of land that hasn't been built on before, you'll need to choose which utility companies you want to use and set up accounts with them. Representatives from each of the companies will come to lay the groundwork so that those utilities will be ready to install when the time comes.

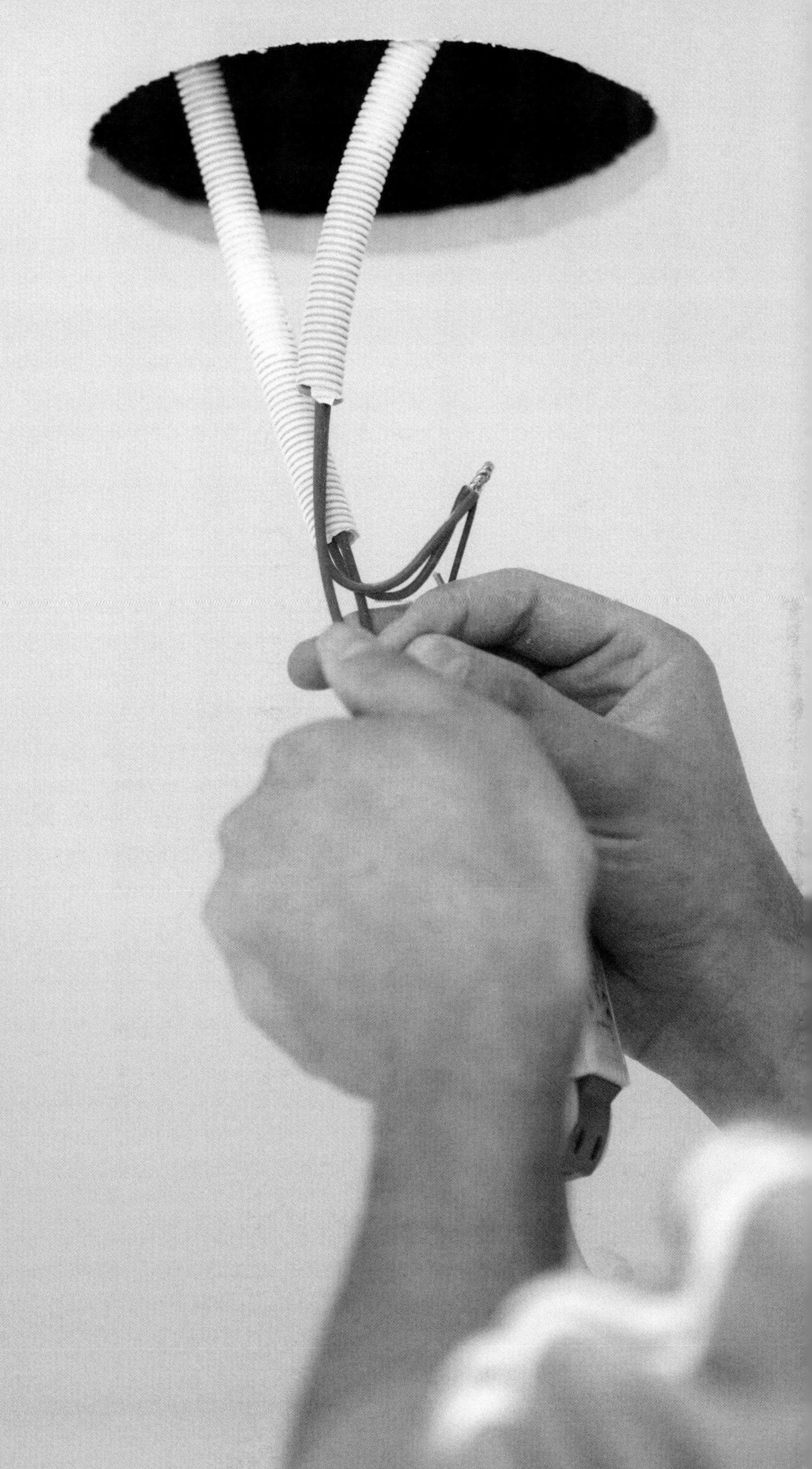

A Word to the Wise

Turn it off

It really is vital that you make sure all existing services have been turned off before any work begins on your property. I was once working on an office, and my foreman told me the whole floor had been deadened – in other words, the switchboard was supposed to be isolated and there should have been no power.

I found out the hard way that my foreman was mistaken. I was sitting on the floor trying to lever off some skirting when I saw a cable I needed to move. Normally, I would have had an electrical tester, which is a little device that tells you if a cable is live or not, but I hadn't thought to bring it with me that day. Thinking the power was dead, I grabbed the cable with my bare hand. The power of the shock threw me about five metres straight back.

I wasn't badly injured but it was a pretty serious shock. It's a feeling you can't really describe. The pain is just incredible – it shoots through your whole body. You really don't want electric shock and neither do any of your tradesmen, so make sure that power is off.

RUNNING A SITE MEETING

Now that you're ready to actually start your project, you should be prepared to conduct – or at least attend – regular site meetings. At each site meeting, you'll discuss your schedule and if anyone is holding up progress. You'll also make sure everyone is on the same page in terms of which tradesman is coming next and you can discuss any changes to the schedule or design that need to be made.

You will have already planned your design long ago, but as you're building, the interior and exterior of the house will start to take shape. You might find you want to make some small changes to your design along the way – structural, cosmetic or both.

You'll also want to discuss your budget with your foreman at every step along the way. You may find at times that your budget won't allow you to achieve the finish you originally planned, so you might have to look at cheaper alternatives as you go. For instance, you might have planned to do your garage in solid brick but you realise you're running out of money so you need to go with a less expensive option.

You will have already drawn up a budget but you'll always encounter little unforeseen problems that you weren't expecting. At each site meeting, your foreman will make those problems apparent and then it's up to you to decide how you're going to fix those problems with the budget you've got.

If you don't feel competent to conduct a site meeting, it's okay to have your foreman do it for you, but it's a good idea for you to be there as often as you can. How much you rely on your foreman depends on you, but the more involved you can be, the better. You have the right to sit in on every meeting between your foreman and your tradesmen, and I recommend that you do.

This is when your people management skills get a real work out, and only you know how capable you are at dealing with people. The site meeting is a good time for you to raise any questions or concerns you might have, but if you can't talk to your employees without offending them, it's better to leave those conversations to your foreman. You may also want your foreman to help you clarify ideas you may have had a hard time explaining.

Making all the big decisions is ultimately down to you. It's your foreman's job to make you aware of the problems and suggest two or three options of how you can fix them, but the choices are going to be yours to make.

A Word to the Wise

Be firm but flexible

You want to be flexible in dealing with your tradesmen, but there are a couple of traits you really shouldn't put up with.

One is lateness. You might be working with a tradesman who does fantastic work but tends to be late in getting the job done by a few days or a week. As the project manager, you can't let that happen. You need to make it clear to everyone that they need to finish their jobs according to the time estimate they gave you, not when it's convenient for them.

Why? When one tradesman runs late, it has a knock-on effect. Everything has to happen in the correct order so if

one person is behind schedule, he or she can throw off the next person, which will throw off the tradesman who follows. Before you know it, you've gone way beyond your original master schedule and that time will start to get expensive.

Also, don't feel like you need to spoil your tradesmen. You can get them a cup of coffee but don't feel like you have to drive down to a café or make them lunch every day. These are your employees, and you're paying them a good wage, so they should be looking after those things themselves. And if you find someone is being rude or sexist or otherwise inappropriate, then you should say something. Even better, have your foreman speak to them.

A Word to the Wise

Be camera ready

It's a good idea to take lots of photos along the way throughout your build, as detailed as you can. That way, if there's ever a dispute, if you're not sure about some aspect of the build or if you need to verify something, you've got evidence. Also, when your project is completely finished, it's kind of cool to put together a slideshow of photos and see how your project progressed from beginning to end.

DEMOLITION

If you're doing a renovation or knocking down a house, you'll need to start with demolition. How you demolish the standing structure depends on a couple of factors, namely whether you're starting over completely or you're saving an existing subfloor.

Starting fresh

If you don't plan to save the foundation or subfloor, a complete demolition usually only takes a day or two. In this case, you'll bring in an excavator to bowl over the house then lift up all the concrete and take away all the rubble as part of the contract.

Saving the subfloor

If you're demolishing your house, there's a remote chance that your foundation or subfloor might be reusable for your design. If you decide to keep the foundation, start by removing the floor coverings and get them back to bearers and joists. We'll talk more about subfloors in Chapter Nine.

Finding a gem

In doing your demolition, you might pull up carpet and discover that you've got some beautiful old floorboards that you might want to reuse and it makes sense to do so. If that happens, simply protect those floorboards by covering them up during the build. Then, at the end, put the boards back in place and get them sanded and polished. Make sure there's no termite, borer or moisture damage. If there is, it's just a simple matter of replacing the damaged boards.

Keeping the foundation

If you want to keep the foundation, you can't just knock down everything. Instead you have to take apart the structure one piece at a time. Otherwise, there will be too much vibration and it will damage your foundation. You can demolish the whole thing by yourself – I did when I built my house – but you need to know what you're doing. If you don't, you can get hurt.

Also, if you're trying to preserve your foundation and floors, your existing walls, roof and floor will have to coincide with your design for your new walls, roof and floor. For this reason, I would recommend using your carpenter to do your demolition if your house is coming apart bit by bit. A qualified carpenter will be able to demolish the old structure while at the same time thinking about how the new structure is going to be rebuilt – a demolition crew might not.

The carpenter can bring in extra people to help with the demolition depending on your budget and time frame. If you need the project to go faster, you can hire extra people to help with the demolition. If cost is a bigger issue, though, you can have the carpenter do it alone or with your help to save money on labouring costs.

A Word to the Wise

Take out the rubbish

In demolishing houses, we have found some really amazing things. For a start, a standard contract for a renovation usually includes a clause that stipulates that whatever the tradesmen find during the demolition is theirs to keep. I've come into some really interesting stuff that way.

I remember once a client forgot about an ancient wine cellar he had and as the demolition team was working, we came across all these dusty old bottles of Penfolds wines. At first we didn't touch them but then we asked the builder what we needed to do with them and he said we were free to take them. We ended up taking that wine home and I sold my share, making a profit of about $6000.

Also, people always want new appliances when they're doing a renovation so a builder or foreman will often find an old dishwasher or refrigerator when doing the demolition. A lot of times, we can reuse those appliances or sell them – which is a great little perk.

Probably the most interesting was a renovation I did at a hotel in a southeast suburb of Melbourne. Before we did the renovation, the place had been well known for being the type of hotel where you could hire the rooms by the hour. All kinds of dodgy stuff had been going on there for the previous twenty or thirty years and we found many, many, many things that we didn't want to find. We'd pull apart bedheads and find sex toys, drugs, money, syringes, pornography – you name it. It was shocking.

Banks are the best. When a team renovates a bank, we can be almost certain to find cash, and if we find it, we can keep it. As we pull apart the counters, we just find cash everywhere, thousands of dollars. I've seen it a few times. You'll normally find that when it's happening, people know about it and they all turn up that day – the tradies will come out of the woodwork to help demolish a bank.

Q&A

What else can I salvage?

If you're going to the trouble of dismantling an old house piece by piece, it's worth looking to see if there's anything else you can salvage. In addition to floorboards, you may also be able to save doors, wall sconces, light fixtures and other fittings that you may be able to use again.

Take a good look around and see what's there then inspect each item carefully for damage. Make sure you wrap or store anything you save to protect it from the weather. If you don't end up using the bits you've salvaged, you can try selling them on eBay, Gumtree or at antiques outlets. Or you can donate them or just throw them away.

LAYING THE FOUNDATION

Even with renovations, the foundation is the first step. The type of foundation you use depends on the type of soil you have. You will have decided long before now what type of foundation you want based on the type of build you're doing, the type of soil you have and the rest of your engineer's report.

Renovations

Before you change or add anything to your existing structure, you need to make sure your foundation will be able to support your new design. This is where your architect and engineer come in. If you haven't already had a soil report, you may need one to determine if or how you can reinforce your foundation. If you are renovating and beefing up your **underpinning**, be aware that you're likely to get some cracks in your plastering, but that's common.

Underpinning refers to a foundation of stumps that are already in place.

Building a new slab

The first step in putting in a slab is excavating which involves digging all of your perimeter **footings** and internal footings. Footings are normally around 600mm deep and 400mm wide and your house sits on that solid layer of concrete.

Footings are the structural base of your house – what prevents your house from shifting or moving. Your foundation will sit on top of these footings.

With this step, a tradesman - either a concreter or a specialist excavator - will come with a little Bobcat excavator and dig out the footings, preparing a flat surface for your slab to sit on. Next he'll install a waterproof plastic membrane followed by **structural mesh**.

At this point, the building inspector will come and look over the application of the steel, the depth of all the footings and the thickness of the slab. Once you get the inspector's approval, you're ready for pouring. A concreter will pour the slab, **screed** the slab and then finish it off at night. Pouring the slab is normally a one-day process although big slabs can take three or four days.

Once poured, the slab will need a few days - sometimes as much as a week - to cure before you can continue on to framing. There's probably no need for further inspection before going on to the framing stage. All the inspector will look at is the depth of the excavation and the actual structural steel that you have in place.

Structural mesh – also called trench mesh or rebar – is a grid of metal designed to give the concrete of your foundation a spine.

A **screed** is a short length of wood or aluminium your concreter will use to smooth out the concrete so that your slab will dry as a flat, level surface.

Building new stumps

With stumps, you'll also begin by excavating. As with a slab, a tradesman will come with a small excavator and usually dig all of your stump holes in a day if you have good ground.

When doing your stumps, make sure they are built to the requirements on your plans. Your engineer will determine the depth, the diameter and the spacing between each stump, and those specifications have to be followed exactly. If you are 5mm over, the building inspector will come along and say, 'These spaces are too big – start again.'

Once the holes are dug, the inspector will come and check to make sure you're okay to have the concrete for the stumps poured. This inspection is your first and definitely a vital one. If your stumps are too far apart, your whole house will be unstable, so you want this inspection to make sure your foundation will do its job.

Once the holes are dug, get the concrete poured straightaway – your carpenter should be able to do that job. You just need the right excavation and a ball around the bottom of each stump. As with a slab, your stumps will need to sit untouched for several days to cure before you continue on with your build.

Building brick piers

Brick piers are built similar to stumps. With brick piers, you'll start by excavating for a concrete pad usually roughly 500mm deep. Once the hole is dug then you'll pour and screed the concrete to a flat surface onto which a bricklayer will build the brick piers. These will be held in place with a layer of mortar and, as with stumps, they'll need to cure for at least a day or two. You'll also need to have your piers inspected.

Whether you're using a slab, stumps or brick piers, allow about a week for putting in your foundation if you're building from the ground up – it could take more for renovations – and include enough time for curing.

Q&A

What the best time for pouring a foundation?

The best time to do foundations is during the dry season. If you dig the footings for a slab and it rains before you get the concrete poured, you'll have one giant swamp where your excavation used to be.

Before you get started, it's a good idea to look at the weather forecast for the upcoming week. If it's going to be terrible weather, you're simply wasting your time trying to do excavating. Hold off until you have a window of three or four days of good weather and then do your foundation.

If you dig stump holes and then it rains, those holes will fill up with water. A hole that's 900mm or a meter deep will fill with water and then you've got dozens of holes full of mud. Also, a stump hole that started off being 400mm wide is now 600mm wide, which means you need twice the amount of concrete to fill it.

Whether you're using a slab or stumps, if you get caught in the rain during the foundation stage, you're in trouble. It will take at least two days to get rid of all of that water, and you have to be sure those footings are completely dry before the concrete goes in. A process that might only need three or four days might turn into three or four weeks. In other words, it's better to do your foundation during the dry season.

EXCAVATING FOR OUTDOOR STRUCTURES

If you're planning on putting in a swimming pool, now is the best time to dig the hole where that pool will go. Unless you have a huge block of land with heaps of access, this step needs to happen now because excavating for an in-ground pool requires lots of big machinery. Once your house is built, you'll probably have a much harder time getting that machinery in. You're also likely to have up to ten truckloads of soil left over from digging the hole for the pool, so you need to have a plan for either using or removing that excess.

The best method is to dig the hole and then pour the concrete for your pool at the same time as you're excavating and pouring concrete for the foundation of your house. Also, you need to have the concrete in now because if you just have a big empty hole in the dirt, the first time it rains, your hole will collapse and lose it's original shape and size.

Once you get your pool to the concrete stage and there's just a shell, you'll need to cover it over because you can't have a huge hole in your backyard for safety reasons. I normally cover over open pools with temporary floor joists and then sheeting or tarp to keep the hole from filling up with rain or anyone from falling in.

If you have any other outdoor structures, their foundations may go in now as well. Your architect will have included any additional foundations in your plans. If you're only just now deciding you might want to include an additional structure, this is a great time to get those foundations in place.

WHY YOU SHOULDN'T DO THE FOUNDATION YOURSELF

The foundation stage is one in which most people shouldn't get involved. Only a very skilled person who knows what they're doing should carry out the foundation work because it's quite difficult to establish a perfect height for your whole house or renovation area.

The foundation is probably the most important thing in your house – it has to be perfect, and you can't cut corners. You need to make sure the person doing your foundation has followed the drawings. They should be familiar with having inspections and they should be able to guarantee the job is done perfectly.

If this job isn't done properly, your entire house will begin to shift from the very beginning, and that flaw is difficult and costly to repair. Your floor will creak and cracks will appear in your walls and tiles. Unless you are a builder or a carpenter, don't even try to do the foundation by yourself.

A Word to the Wise

Get involved

You shouldn't help with the technical work involved in pouring a foundation, but that doesn't mean you shouldn't do anything at all. One good reason you should get in there and help is so that you can keep an eye on everything that's happening at the site. Also, you get to build a rapport with your tradesmen and that can sometimes translate into better workmanship.

Manage with kindness

Once your tradesmen are on site, you're the boss. I've heard the expression 'you catch more flies with honey than you do with vinegar', and it's true for managing your tradesmen, too.

I once worked for a builder for seven or eight years as his foreman and one time, I made a mistake – I forgot to put **noggins** in the walls to support some shower screens. In all the time I'd worked for the guy, it was the only time

I ever stuffed up but he started abusing me over it, so I just left him.

Mistreat your tradesmen or manage them badly and you may find yourself building your house on your own.

Noggins are the small pieces of timber that sit horizontally between the vertical studs in the frame of a wall.

How can I help?

Even if you aren't a skilled foundation specialist, you can still help out at this stage. This is when you'll want to listen to the experts and follow their directions. If you don't, you could end up hurting yourself or someone else quite badly. You might also make a critical mistake that could cost you a lot of money to fix.

Demolition

You can help with the demolition under the direction of your foreman. If you don't feel comfortable doing demolition work, you can always pick up bits of rubble and safely put all of those discarded materials into bins.

Clearing stump holes

Whenever the excavator is using the digger, he or she won't be able to get into each corner. You might be able to grab a shovel and neatened up those stump holes by hand. Just be careful – if you get carried away, you might end up making the holes too big, in which case you won't pass inspection. Ask the foreman or excavator to show you how to do the job and then have him or her come back and inspect your work periodically to make sure you're doing it properly.

Cleaning up

Throughout your build, you'll be able to get in there to keep the site clean. Your tradesmen will expect a nice, clean work site and if it's too messy, they may feel it's unsafe and refuse to start their contracts. There will always be packing materials like empty concrete bags or other rubbish you can pick up and take to the bin. If you're digging stumps, there will be an incredible amount of dirt everywhere. Some of it will be used to help reinforce the concrete and you may be able to reuse some of it for your landscaping. Ask your foreman where you can put dirt to save or reuse. The rest you can put in your skip.

Labouring

You'll save yourself a lot of money by doing as much of the labouring as you can. There will be plenty of jobs to do that don't require a lot of skill, things like moving materials to the ideal place or handing tools to the tradesmen, even making cups of tea. Just show up and ask the foreman or tradesman on site what you can do. If you're paying someone else to do it, it's going to cost you $250 or $300 a day – multiply that figure by five days a week and it adds up fast.

Ordering concrete

One thing you should know is that the building inspector doesn't actually watch the concrete for your foundation being poured. All your inspector has to do is look at those holes in the ground and say, 'Right, you're good for concrete.' The inspector doesn't watch the concrete go in and doesn't see what type of concrete is being used.

As the owner builder, you can order the concrete yourself so you know it's the right specification. If you don't know much about concrete, that's okay. As with the rest of your build, you can get direction from your concreter or your carpenter and then you can order it yourself – the engineer should tell you how much concrete you need and what type.

A Word to the Wise

Use proper lifting techniques

There's a right way and a wrong way to lift. Before you lift anything, stop and think about where you're going to put it. Keep the object close to your body if you can and place your feet about shoulder width apart. Keeping your back straight, bend from your knees and lift with your legs. If you feel yourself beginning to struggle, stop and ask for help or lighten the load. Wheelbarrows can be especially dangerous if they're too heavy, so be careful not to overfill them. If you need to, use a belt or back support – if you're not sure, ask you doctor if you should wear one.

CHAPTER SUMMARY

The first thing you need to do is get your job site ready. You'll need someone onsite at all times with a first-aid qualification and you must have the proper insurance. You'll also need toilets and other amenities for your tradesmen, a site office (maybe), access to your property, temporary water and power, fixed ladders, handrails and security fences. Also, always put safety first.

Whenever you go on the job site, make sure you're wearing your PPE – personal protective equipment, that is – including boots, hard hat, gloves, and protective glasses.

One job you can do yourself is cleaning up. There will always be heaps of rubbish on site, so you'll want to organise a skip. Depending on what you're building or renovating, you may need to have your skip emptied several times before your project is finished.

Now is the time to sort out utilities including gas, water, electricity, sewerage, heating, cooling, telephone/Internet and whatever else you may need once your house is finished. If you're doing a renovation or demolition, make sure all services have been turned off before work begins. If you're starting with an undeveloped plot of land, you'll need to set up accounts with each service in anticipation of your build.

Throughout your project, you'll need to conduct (or at least try to attend) weekly site meetings. At these meetings, you'll discuss your schedule and budget plus any changes you want to make. You can have your foreman help you or even run these meetings for you if you want.

Take lots of photos throughout your build from beginning to end. It will help you make a case in the event of a dispute, and it will create an interesting keepsake.

If you're demolishing, you can bring in an excavator and bowl over the entire structure including the foundation – or if you want to save the subfloor, you can dismantle the house piece by piece.

If you decide to save your floorboards, pull them up at the demolition stage, inspect them and then protect them from the weather until you're ready to reinstall them at the end of the project.

You can also salvage other features, fittings and materials from an old house to use again, sell or donate.

Once demolition is finished, you can begin laying your foundation. With a renovation, you may need to reinforce your existing foundation.

With a new slab, you'll dig footings, line them, reinforce them and then have them inspected – upon passing inspection, you can pour the concrete and let it cure. With new stumps, you'll dig the stump holes, clear them and have them inspected. Once you pass inspection, you can pour and cure the concrete. With brick piers, you'll dig footings and pour the concrete pad then build the piers on top of the pad.

If you're putting in a swimming pool or other outdoor structures, you'll probably want to do the excavating and pour the concrete at the same time that you excavate and pour the concrete for your foundation. Just make sure you cover the hole for safety.

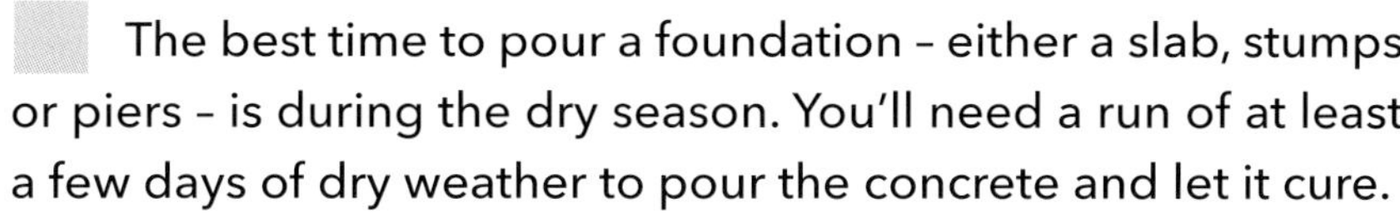

The best time to pour a foundation – either a slab, stumps or piers – is during the dry season. You'll need a run of at least a few days of dry weather to pour the concrete and let it cure.

You really shouldn't do the foundation yourself. It is a very detailed job that's hard to get right and can cause real problems if you get it wrong.

It's a good idea for you to stay involved with your build as much as you can. Remember that, as the owner builder, the responsibility lies with you.

There's a lot you can't do – and shouldn't try – but there are some jobs you can do, including demolition, clearing stump holes, cleaning up and ordering concrete.

You can do yourself serious damage if you don't follow the proper guidelines for lifting, including proper stance and lifting with the legs.

WHAT COMES NEXT

Your foundation will need to cure for a few days or a week depending on the type of concrete you've used and the weather – your engineer or concreter should be able to tell you how much time to allow. Once the foundation has cured properly, it's time for framing.

CHAPTER NINE

THE FRAMING AND THE LOCK-UP STAGES

Now that you have a solid foundation, you're ready to put up the frame and the roof and get your house locked up. By the end of this stage, your structure will have really taken shape and you'll be able to get a very good idea of what your new house will look like.

ANT CAPS

Whether you live in an area prone to termites or not, I recommend that you install **ant caps** before continuing on. Usually made from galvanised sheet metal, the ant caps go in between the foundation and the timber of your floor. Your engineer will tell you if this step is necessary, but I would urge you to do it regardless. You never know when termites might be introduced to your property by accident.

Ant caps, usually made from galvanised sheet metal, are designed to keep termites from getting into your subfloor.

A Word to the Wise

Watch the neighbours

You're probably pretty excited about building your house, but your neighbours may not share your enthusiasm. If you find your neighbours do not like you, your project or your employees, then you may need to intervene between them and your tradesmen.

I remember having to deal with a particularly difficult neighbour around fifteen years ago. I was working as a carpenter, doing an extension on a house in Port Melbourne for the client of a friend.

The lady who lived next door to the build was really grumpy. I was working alongside another carpenter and we didn't get off on a very good foot with the neighbour. Early on in the job, before we knew she was there, she overheard us swearing one day and she asked us to stop. We did, but things just deteriorated from there.

She was constantly watching over us, abusing us and complaining about everything we did. She didn't like where we parked our cars (even though we never blocked her in), she didn't like the mess we were making (even though we cleaned it up) and she didn't like the noise (even though we were working in the middle of the day). Basically, she just wasn't very neighbourly.

One day she kept on and on giving us grief, and my mate and I decided we had finally had enough. We were putting up the frame of the house and it was stinking hot, something like 40 degrees. We decided we needed to cool down, so we stripped off and continued on working – wearing absolutely nothing but our nail bags.

Well, the lady next door just about lost her mind and about half an hour later, the cops showed up. They took one look, burst out laughing, told us to put our pants back on and left. We probably didn't handle the situation in the most professional way, but we put our pants back on and carried on working – and the lady next door never said another word to us.

THE SUBFLOOR

There are a couple of ways that the subfloor can be built: over stumps, over brick piers or over a slab.

Building the subfloor over stumps

If you're using stumps or piers, you'll need to start by putting in bearers and joists to establish your floor structure. It's called your subfloor framing. If you have a concrete slab, you'll skip this step because the slab will serve as your subfloor.

Putting in the bearers and joists

You'll start by putting in the bearers. Bearers are normally 90x90mm hardwood timber or LVLs, which are just two pieces of 90x45mm that have been laminated together to create a thickness of 90x90mm. Your bearers will sit directly on top of your stumps to create a solid base for your house.

Normally the stumps will be placed with a steel spike coming out the top of each one. Your carpenter will then drill a hole through your bearers and fit them over the spikes in the stumps so that the spikes protrude through the holes. The carpenter will then bend over each spike to lock the bearer (and thus the entire floor) to the stumps. As you can imagine, this is a very important step.

Now come the joists. Joists are normally 90x45mm hardwood – they can be made of engineered wood, steel or concrete, but hardwood joists are the most common. These will lay perpendicular on top of the bearers directly beneath them. The joists are closer together than the bearers and create a sturdy grid on top of which you'll lay your floor covering.

Putting in the floor covering

Once you've installed your bearers and joists, you can put down your floor covering. **Yellow tongue** flooring – also known as chipboard or particleboard flooring – is the most common flooring material but you can always put down floorboards or lay carpet over the top of the yellow tongue.

Yellow tongue – also known as chipboard or particleboard – is the most common flooring material. It's not very attractive, but it's water resistant and usually pretty cheap.

In areas where you plan to use tile as your floor covering, you'll use heavy compressed sheet instead of the yellow tongue as the structure for the underside of the tile. You can also lay yellow tongue and then install tile underlay, but I would recommend compressed sheet. Again, this step is unnecessary if you're using a slab instead of stumps.

You may want to install nice quality floorboards on top of your yellow tongue, but if you do, that step should wait until the very end of your project - otherwise your floor may get damaged during the build. Walking on top of floor joists can be quite dangerous, though, so it's a good idea put down something temporary like plywood or yellow tongue to give you and the rest of your tradesmen a flat, safe working area while you're standing up your walls.

It's imperative that your subfloor is built flawlessly because heights with floors have to have millimetre perfect - you don't want to have any steps between rooms or from a passage to a toilet or bedroom. It needs to be one flat plane. It's up to your carpenter to determine the height of your floors, and to make sure your subfloor is level using a laser level or **theodolites**.

Theodolites are highly precise instruments that make exact horizontal and vertical measurements, mainly for the purpose of establishing that a surface is level.

ACCOUNTING FOR FALL

When you're building the floor of a bathroom or a laundry, you'll need to have falls in your floor. A **fall** is simply a tiny amount of slope that allows water to drain. That way, if your washing machine leaks or your tub overflows, the water will be able to escape.

Without falls and drains, even a small overflow of water can flood your house and cause $20,000 or $30,000 worth of damage. You won't have falls and drains in every room, but you'll always have them in wet rooms and on a first floor.

A **fall** is the tiny amount of slope in the floor of a wet room or the first floor of a house that allows water to drain.

A Word to the Wise

Don't have the expertise? Don't touch the subfloor!

Unless you really know what you're doing, you probably shouldn't get involved in building the subfloor. You don't want a floor that's out by more than 5mm, and that's a pretty small tolerance. You want your floor to be perfectly level throughout. To make that happen, you need to have experience using laser levels or theodolites. If you don't have the requisite experience or expertise, don't even try – you could end up ruining your house.

BUILDING THE SUBFLOOR OVER A SLAB

If you've got a slab, you won't need bearers and joists but you'll still need some kind of floor on top of the concrete. You have a few options.

Floating floor

A **floating floor** sits directly on top of your slab. In contrast to floorboards, a floating floor is usually made of vinyl or laminate and it isn't attached to the subfloor. Basically, it just fits together in pieces so it's very easy to install and one of the most inexpensive floor coverings you can use. There are a lot of very nice looking options for floating floors, but I personally prefer polished floorboards myself. If you're going to install a floating floor, just leave the foundation or subfloor for now and put in the floating floor at the very end of your project.

A **floating floor**, which is usually made of vinyl or laminate, sits directly on top of your subfloor or slab and fits together in pieces.

Battens and floorboards

If you want proper floorboards, you'll need to attach floor **battens** to the slab.

Battens are long flat strips of timber or metal. In a floor, they attach to the foundation to create a surface on top of which to lay floorboards. You'll also have battens in your roof – they create the surface on which your tiles or steel sheet will sit.

Floor battens are generally 90x35mm or 90x40mm lengths of hardwood or pine and they serve as a type of mini joist. When you're ready, you'll then lay your nicer quality floorboards on top of the battens, but not till the very end.

Insulation and carpet

A lot of people prefer carpets in bedrooms and sometimes throughout the house. It's a good idea to lay down a layer of insulation and then the carpet, but you won't do any of it yet – wait till the very end to lay carpet.

Temporary flooring

At this stage, you probably don't want to install your nice floorboards or other covering yet – most of the building is yet to come and between weather and muddy boots, you don't want your floor to get wrecked. A good alternative is to put down a layer of yellow tongue as a temporary floor covering because it can stand up to the weather for two or three months if need be. Another temporary material is plywood, which will also provide a nice even floor for everyone to use during the remainder of the build.

A Word to the Wise

Watch your head

I cannot stress enough how serious safety on the job site is. I remember once I was working at a private school building a big glass atrium. This atrium was huge, about 50m long and 25m wide, and it enclosed the entire courtyard of the school.

I was working on putting up a big triangular section of the frame that went out in the very front of the structure. It was awkward to manage, so instead of putting up the frame directly, I took measurements and then built that section of the frame on the ground.

When it came time to install the triangular section, I hauled it up on a scissor lift. The problem was that the bottom of the frame was probably 15m off the ground and the scissor lift didn't reach. I was in a hurry so I didn't take the time to find a safe solution. Instead, like an idiot, I got out of the scissor lift and climbed up onto the frame.

I was able to slide the triangular section into the rest of the frame, but it was a tight fit, so I decided to use a hammer to bang it into place. The only way I could reach the spot I needed was to stand on the steel frame, hanging on with one hand and swinging the hammer back towards my own face with the other.

It may come as no surprise that eventually I missed a swing - which happens a lot in carpentry - but because I was hammering backwards, I managed to ping myself directly between my eyes with a full swing. Immediately, blood started gushing straight out of my forehead like a fountain. I looked down at my mate who was still in the scissor lift and by the look on his face, I could tell he thought I was going to die.

If I had knocked myself unconscious, I would have fallen and almost certainly killed myself. As it was, I went to hospital, got five stitches and took the next day off and the following day I was back at work. The worst was that I was around 30 years old - old enough to know better - but I learned my lesson that day and never did anything like that again.

FRAMING

Once you have a viable foundation and subfloor with at least a temporary floor covering, you can start putting up the walls and the frame for the roof. There are a few different ways you can go about framing your house.

Prefab

Probably the easiest - and one of the most cost effective - ways of framing a house these days is using **prefab**, or prefabricated walls. Start by consulting a few timber truss companies, a steel framing company such as Supaloc or even a large hardware store like Mitre 10. Based on your plans, they'll use an AutoCAD program or something similar to design and build all of your walls and your roof structure in a factory.

Prefab – short for prefabricated – refers to a method of framing whereby the frame of the walls or roof are designed by a program like AutoCAD and then built and assembled in a factory.

Once the panels are delivered, all your carpenter needs to do is stand up those walls and pop on that roof - putting them together like Lego. With the average house, you can have most of your prefab walls stood up in one day. Then the next day you come along and put up the roof.

This method is good because, whether you're using timber or steel, the parts are constructed in bulk in a factory and it takes very little time to install. Also, a computer calculates the measurements so they're very nearly perfect.

Stick build

Your carpenter may choose not to go prefab. Some homes are very intricate and they may not allow prefab, especially if you're doing a renovation or replacing part of a wall. In that case, it may be more feasible to have the carpenter **stick build** the frame - in other words, put together the timber to form each section of the frame one piece at a time.

A **stick built** frame for a wall or roof is constructed piece by piece by a carpenter.

If you're just doing two rooms and a bathroom and the framing component is going to be stick built, it should only take a couple of days. If you're going to build a whole house manually, you may need to allow more like a week for the frame and another week for the roof.

Solid brick

There are a lot of reasons you might want to use brick for framing your walls including durability and insulation. With a **solid brick** wall, there's no timber – both the frame and the cladding are brick. While durable and pest-proof, solid brick walls are expensive and they take a lot more time to construct than a simple frame, usually a few weeks. With this method, a bricklayer does all of the work in conjunction with your carpenter. Together, these two tradesmen will work out the exact spacings for your windows and your doors.

A **solid brick** frame is composed of a brick frame plus a second layer of brick veneer.

Other framing materials

If you're using other, less conventional building materials like straw or mud bricks, your framing method and time will vary. Some benefits of using alternative materials may

be that they save you time and money, or they might be more environmentally sustainable but take more time than some of the traditional framing methods.

Regardless of what you use, your foreman and other tradesmen should be able to give you a fairly accurate estimate of the cost and the time they'll need when you draw up your contracts before the build begins.

THE FRAME INSPECTION

Now that the walls are standing and the roof is on, your house is really starting to take shape. Before you begin cladding, you'll need to get a frame inspection.

This is a critical check - if you start cladding without the inspector's stamp of approval, you'll have to pull apart your premature work so that the inspector can see that frame.

You may need another inspection at the end of the lock-up stage depending on what you're doing, the materials you're using and your local council's building regulations - but in many cases, the frame inspection will be the last until your final check.

You'll still need to have the building inspector come and approve your build for the Certificate of Occupancy, but usually you won't have another inspection between now and the very end.

BUILDING THE ROOF

As soon as your frame is up, you want to get your roof on as quickly as possible. Once the roof goes on, you'll have a protected work area so right now, speed is essential. The fastest way to get a roof up is to put steel sheeting over a prefabricated trussed frame. A **trussed roof frame**, like the frame of the walls, creates a grid of timber on top of which to lay your roofing materials.

Your carpenter or foreman may deem that prefab trusses aren't the best option for your build. If you are doing it by hand, allow maybe a week. Otherwise, you should be able to get the roof frame up in just a couple of days.

Once the frame is up, you'll need to put up a layer of **sarking**. Sarking is a fancy word that just means the process of putting up foil, plastic or paper insulation over a truss frame so that you have an extra weatherproof layer between the frame and the roof cover.

A **trussed roof frame** is designed to create a grid of timber on top of which to lay your roofing materials – the truss is one of the supporting legs of the frame.

Sarking is insulation that goes on both roofs and some types of walls.

Once the sarking is up, you're ready for your roof cover. You have a few options, but in Australia the three most common types of roof cover are tiles, steel sheeting and (to a lesser degree) shingles. We talked about a few of the pros and cons of each in Chapter One - here's a quick look at installation.

Tiles

Many people like the strength and durability of roof tiles. If you're going with roof tiles, keep in mind that they weigh ten times what a steel sheet will weigh. For this reason, you'll need to allow more time to install tiles than most of the alternatives. How long will depend on the size of your roof.

Steel sheeting

You might prefer the look and durability of Colorbond, or that may not be as important to you. While very hearty, steel sheeting is somewhat less sturdy than tiles, but they're quick and easy to install – plus they make that fantastic noise in a heavy rain!

Shingles

Though not as common in Australia, some people prefer the look of shingles, which are widely used in America. Though not as heavy as tiles, shingles must be installed one at a time so they'll take more time than steel sheeting but less time than tiles.

There are different methods of building depending on where you live and what environmental factors are likely to influence your house. Consider this overview as a guideline – every build is different.

A Word to the Wise

Watch your step

It's up to you to police safety on your job site. Falls are common, and they can be quite serious. I had a friend who was putting up ceiling joists once and he fell from about five metres. He was just climbing around on the roof without any fall protection, which is never a good idea.

As he fell, he landed on one of the floor joists and snapped the bone in his shin cleanly. His leg has never really repaired. If you see someone doing something that's unsafe on your job site, like climbing up a ladder and standing on the very top stair or not wearing PPE, stop him or her. If the person falls and gets injured, it's on you as the owner builder to compensate him or her for worker injuries. All health and safety falls on your shoulders so make sure your employees are working the right way.

STRAIGHTENING

You've got your frame up and your roof on, so now you want to make sure it's all plumb, straight and level. This essential part of the build is called straightening.

Once a frame goes up, the timber will never be perfect. It will almost always have bows and other imperfections, so it's imperative that you have a skilled carpenter straighten your frame. Otherwise you'll have curved walls that will show out on all of your intersections.

At this stage, your carpenter needs to go over the whole house and make sure all the walls are straight and everything is fixed off properly. This step absolutely needs to happen now because once you've covered up your frame with cladding and plaster, there's no going back.

WINDOWS AND EXTERNAL DOORS

Once the roof is on, it's time to install windows and external doors. During this process, your carpenter will make sure all windows and doors have the appropriate **flashings**, which is to say, the proper seals. This is a small step, but it will make a big difference to how well your house is protected from the weather once everything is finished.

> **Flashings** are the seals that go around your windows and external doors. They are essential for protecting your home against the weather.

SARKING THE WALLS

Just like you had to put sarking on the roof before laying the roof covering, you'll probably need to put sarking on your walls as well, especially if you're rendering. The only time you won't do sarking is if you're using double brick.

Love it? Learn it!

There are a lot of jobs that will need doing that don't require a huge amount of skill. Remember that a lot starting apprentices are teenagers straight out of high school, so there's no reason why you can't help.

Having said that, if you don't listen and follow directions, you could stuff up your job and ruin the project. If nothing else, you can clean up and feed materials to the tradesmen.

If there is some aspect of the building work you find especially interesting, however, you can always go and get your own certification in that field. There are TAFE courses everywhere that will teach you what you need to know so that you could do some carpentry, plumbing or electrical work yourself. If you really enjoy it and find that work satisfying, go take the course. You may find yourself taking on a new hobby or even a new career.

In reality, though, you've already contracted all of the materials and labour so you shouldn't really need to do much of the labouring. The time you might be tempted to bite off more than you can chew is if you've had problems and you're running short on money. If that happens, you need to think long and hard before you try to do a job you're not really qualified to do well.

RUNNING SERVICES TO EXTERIOR STRUCTURES

Prior to cladding, you need to make sure you have power, plumbing and gas in place for all of your exterior structures. If you're building a freestanding office or granny flat, you may also want to run cables for phone or the Internet as well. Normally everything runs from a switchboard that will be located in the front of the house. If you don't have those cables in position prior to cladding, you'll have to pull off some plaster inside the house, get the cables to your switchboard and then run them outside, and that could cost you thousands of dollars.

You may also want to run gas lines to your BBQ. Rather than using gas bottles, have a direct line. It's a cheaper way of doing it and you don't have to worry about running out of gas. As long as you do it prior to cladding, you're in good shape.

A Word to the Wise

Plan for the year

Don't just plan for the season you're in when you build. Consider what those outdoor spaces are going to be like all year around.

The design for my house included a covered patio area, but I didn't realise how much I'd like to have a heater out there until after I moved in. If I want to do it now, it will mean pulling off plaster and making a huge, expensive mess. It's better to think about every type of weather you're likely to experience and include all of those services before the cladding goes on.

CLADDING AND LOCKING UP

Locking up means sealing off the façade of the house. The lock-up stage generally involves putting up your cladding so that it overlaps or attaches to your windows and external doors. Once this stage is finished, your house will be locked up against the weather and intruders. The lock-up stage is very important because this is how your house is going to look from the outside.

In general, the lock-up stage should take a few weeks, although in some situations it may be only a few days – it just depends on the materials you're using and the weather. You may not have all of your cables and pipes in place for the interior of your house yet, but as long as you haven't started plastering, you may still be able to start cladding.

There are lots of different types of cladding including brickwork, weatherboards, foam and render, cement sheets or blue board and render. Bricks take the longest – probably a week or two – as they go in one by one. A weatherboard house will take maybe three or four days to clad the whole thing. Foam or blue board will only take maybe three days.

If you like the look of brick but you don't want to spend the time or money on solid brick, you can go with a brick veneer, which is simply one skin of brick and one skin of timber frame. With brick veneer, you start with a timber or steel frame, either prefab or stick built. The bricklayer will then clad the house in a layer of bricks and tie the bricks to your frame with brick ties. This is the most common type of cladding used in Melbourne and Sydney.

With your cladding, your windows and your external doors in place, you've got your house locked up. As with the roof, you should strive to get your lock up done as quickly as possible – it protects the house from the weather, it stops people from getting into your house and it gives the tradesmen a sheltered area to work in.

EAVES AND GUTTERS

The **eaves** of the house allow water to run away from your house rather than down your walls. Eaves also help to create shade and keep a house cool in summer months. Part of the cladding process includes putting up **fascia boards** or cement sheet to protect the eaves and maximise the overhang. You'll also attach your guttering along your eaves, so if you're installing a water tank (and you should) you'll put up your gutters now.

The **eaves** of the house are the bottom bit of the roof that hangs over the tops of the walls. They allow water to run away from your house rather than down the walls.

Fascia boards are the long, straight boards that run along the lower edge of the roof – they usually support the bottom row of your roof tiles and your gutters are usually attached to them.

RUNNING CABLES AND PIPES INSIDE

Now that your frame is up, you need to put in all the cables and pipes you'll need for your utilities. These elements usually go in after cladding but before plastering. You should have already organised your utilities with your providers, so now is the time to get all the pipes and wires you'll need for heating, cooling, electricity, gas, water, sewerage, telephone/Internet and whatever else you may plan to use.

If you're planning to install solar panels – and I recommend that you do – you need to get a representative from the solar company to come now. Usually, those solar panels will need to be connected to your water pipes and your main switchboard, and the best option is to run those connections through the frame.

It may be a temptation to wait to install solar panels until after the house is built, but if you wait too long, it's highly likely your solar installation technician is going to have to pull off your plaster to install the system. For that reason, you're better off getting the initial groundwork for the solar system installed at the same time as the rest of the utilities.

MAKING CHANGES

Now that your house is starting to take shape, you might decide to make a few changes - make a window bigger, shift a wall or swap a window for a door. If your budget has been thrown out of whack, you might want to swap out a brick wall for weatherboard or use a different type of cladding. One of the benefits of being an owner builder is that you have that flexibility.

The cost of making changes at the framing stage is normally only a few thousand dollars at most. Compare that figure to the many thousands of dollars a volume builder would charge you and you'll be glad you're the owner builder.

If you are going to make changes, remember to go back to your architect to get your plans amended. In your final inspection, your inspector will check to see that the house you've built matches the architect's plans. If it doesn't, you won't pass your final check and get your Certificate of Occupancy.

Also, you'd normally pay for materials prior to delivery but if you change your mind, you can still cancel or change your order and get your money back. If you make the change after the materials have been delivered, keep in mind that those materials need to be perfect if you want to send them back. If it's timber, it needs to be stored flat and straight, and it needs to be wrapped in plastic to protect it from weather. If your materials are even slightly damaged, the supplier won't take them back.

Q&A

I'm not sure if this bit looks right or not. Can I get a second opinion?

If you're not entirely certain about some aspect of your build, get a second option from a licensed builder, a carpenter or a building inspector. They can come in for a day at a cost of no more than a few hundred dollars to give it a good looking over. If they spot any flaws, get those issues sorted immediately before proceeding to the next step.

A Word to the Wise

Fix the mistakes

If you discover a problem with your build or you see that something has been done incorrectly, make sure that the problem gets fixed immediately.

I was once putting on roof battens for a huge mansion on the Mornington Peninsula overlooking a cliff. I thought I was finished, but when the plumber came in, ready to install the steel sheeting, he noticed that I had made a mistake – I had only nailed the battens down but I hadn't screwed them. With roof battens, you have to both nail and screw them to meet code to prevent the roof coming away from the battens.

Luckily my plumber noticed the problem and told me so that I could fix it. If he hadn't caught my mistake, that roof would have come off by now.

HOW CAN I HELP?

When it comes to framing, you should probably leave most of the detail work to the professionals. If your measurements are off by even a few millimetres, your windows and doors won't work and it'll cost you thousands to get it fixed. With only 5mm or 10mm of tolerance, there's every chance that you'll stuff it up. There are still jobs you can do, though, namely labouring, cutting and cleaning up.

Labouring

Remember Waz, the country boy who won the 2011 season of *The Block* with his partner Polly? When he first arrived on the job site, he wasn't able to do much, but as we started building, he would get involved. He got in with the carpenters and learned how to use the tools and do the work.

By the end of the twelve weeks, after putting in eight hours a day every day, he was basically a carpenter - not totally, but he was able to frame up walls, be handy and get a lot of the job done. If you can do the same, you can save heaps.

When it comes to labouring, think about these figures - the average labourer will usually make $45 or $50 an hour, which adds up to over $80,000 a year. If your project is going to run for six months, that's a huge amount of money you'll pay for someone who will often be doing simple grunt work that doesn't require any particular skill.

You can always feed materials to your carpenter. If you have a couple of packs of timber dropped at the front of your house, it will normally take two or three hours to get all that timber moved to where you need it to be. You can also hand pieces of timber to your carpenter as and when required – but make sure you follow directions and don't get in the way.

Cutting

There are also lots of simple jobs that most people should be able to learn. One of those is running a drop saw. If your carpenter has opted to stick build your frame rather than using prefab, you'll find that your studs are nearly always the same length and you'll probably have 300 or 400 of those. And because the studs will be covered with cladding and plaster, there's a slightly bigger margin of error.

Normally the carpenter will set up a drop saw with a bench and a stop so that each piece is cut to the same length. All you have to do is put a piece of timber onto the bench, settle it against the stop and cut. If you feel confident in doing it, you can cut all the studs and noggins. If you don't feel comfortable doing it, however – if you're afraid you're going to mess up or cut off your finger – don't touch it.

Insulating

You may be able to help with putting up insulation, as well. You can put in insulation as long as you leave a gap of 200mm around any light fixtures or electrical cables to prevent fires. If you're not completely sure, I would recommend you only help with the project and let the professionals show you what to do. Get this step wrong, and you could end up burning down your house.

Cleaning up

Doing the cleaning up yourself is an excellent way to save a lot of money, especially during the framing and lock-up stage. Framing makes a huge amount of mess. There's sawdust going everywhere and you've got little off cuts all over the place. Just about anyone can get in there and put all that stuff in the bin. There's no reason to pay a tradesman $50 an hour to do it.

A Word to the Wise

Pay attention to details

If you're going to help, it's never a good idea to be overly confident. On one job, I had to install locks on about 150 doors. I was in a hurry so I hired a machine to cut the holes in the timber doors where the locks would sit.

I thought I was working right on schedule, but I didn't check the machine carefully enough. Somehow the safety had come off and I ended up cutting all 150 of those holes the wrong size – they were all too big so none of the locks fit properly to the tune of about 10mm, which is a lot when you're talking about a door lock.

I had to go back to every single one of those doors and add a piece of timber to every hole to make the locks fit. I'd been going quickly trying to save time but that one repair job ended up taking me about three weeks plus several thousand dollars of my own money to fix.

In other words, don't just go hammer and tongs – check everything you do. Working slowly and methodically with lots of attention to detail is the best way to get the result you want.

A Word to the Wise

Place your materials properly to save time and money

It takes a long time to move materials from one place to another, so you want to have the delivery people strategically place those materials so that they're accessible but out of the way of everyone working. You're paying to have those deliveries unloaded, so you might as well kill two birds with one stone.

If you don't know the best place for each material – and you probably won't – consult your foreman. The plumber might need to dig trenches for drainage or sewerage, or your electrician might need to excavate. If you have to move all those materials again, it's going to cost you a couple of hundred dollars every time. Before you allow anyone to take materials off the truck, ask your foreman where each thing should go.

CHAPTER SUMMARY

- Now that you have a solid foundation, you're ready for framing. If you live in an area prone to termites, you'll probably start by installing ant caps. I would suggest you do so even if you don't live in an area with a high risk for termites.

- If you've used stumps, you'll start by building a subfloor, which involves securing parallel bearers to the stumps below them and perpendicular joists on top of the bearers. You won't put on your final floor coverings at this stage, but you will put on a temporary covering and account for fall.

- If you don't have the proper level of expertise, don't touch the subfloor.

- There are a few options for floor coverings, including a floating floor, battens and floorboards or insulation and carpet. You won't install these coverings until the very end of your build. For now, you'll just want to put down a temporary floor.

- Once you have a viable subfloor, you're ready for framing your walls. The fastest and easiest method is to use prefabricated frames, although your frame may be stick built, built from brick or built from some other material.

- At this stage, you MUST pass a frame inspection before moving on to cladding. If you don't, you'll have to pull off your cladding for the inspector.

- As soon as the frame is up, you want to move as quickly as you can to get your roof on. The fastest way is to put up a prefabricated trussed roof frame, although you may need a stick built roof frame.

- Next comes sarking (or insulating) the roof and then putting on the roof covering, most often tiles, steel sheeting or shingles.

- With a frame and a roof in place, your carpenter should take some time to check the entire house and do any straightening that may be necessary.

- Once the frame is straight, you can put in all of the windows and external doors. Make sure the flashing is in place.

- At this point, depending on what materials you're using, you may need to sark your exterior walls.

- Before you begin cladding, make sure all of the necessary pipes and cables are in place running from the main switchboard of the house to all of your exterior structures. You won't turn on any services yet, but the groundwork needs to go in now.

- Now you can begin the lock-up stage, starting with cladding. Cladding involves putting up the external façade of the house. The time and cost involved will depend on what type of cladding material and method you use.

- Part of cladding includes putting up fascia boards over your eaves and installing gutters to protect your walls from water damage.

- You're getting close to time for fitting out – first you need to make sure all of the cables and pipes you'll need for your utilities are in place within the frames of your walls.

You can still make changes at this stage. If you do, the cost will be a lot less for you as an owner builder than it would be if you'd hired someone else to do the job for you. Remember to take your changes to your architect so your plans match your final product. You can also change materials.

If anything looks wrong to you at this stage, it's worth getting a second opinion and getting any flaws fixed.

You can help in the framing and lock-up stage by doing some of the labouring, cutting timber, putting up insulation or cleaning up.

When you have materials delivered, have them dropped off in the ideal spot on your property to minimise the time and effort involved in moving them.

WHAT COMES NEXT

Once lock up is done, the house will be sealed against weather and intruders. From the outside, your house is starting to look like a home, but there's still a lot to do inside. Now we're ready for fitting out and finishing.

CHAPTER TEN

THE FIT-OFF STAGE

We're getting close to the end, but there's still a lot of work to do during the **fit-off** stage before your house is finished.

The **fit-off** stage is when all of your fittings and services go into your house, including things like cabinets, vanities, closets and interior doors plus all of your electricity, plumbing, gas, heating, cooling and telephone/Internet.

A Word to the Wise

Acclimatise you floorboards

If you plan to install proper timber floorboards, they need to be on site before they go in, at least two weeks prior to when you're ready to install them. This allows the timber to expand or contract based on the normal levels of moisture that will consistently be present in your house.

PLAN YOUR POWER AND FITTINGS

By now, every service in the house has been run. You won't have any services turned on yet, but all cables and pipes should be in place for electricity, water, gas, heating, cooling, solar and telephone/Internet. All the power points will be run through the walls but they're still loose.

Before you continue on, take a long slow walk around the house. Think of where your want to put your beds and couches, and then think about how you want to incorporate your power to suit the way you live. It's a personal thing and you get to decide on everything, including the number and height of light switches, the placement of power points and light fixtures plus the best location for power points for air conditioners, refrigerators, ovens and cookers.

You'll need to make those decisions now, though, before you start putting on plaster, because that's when all of those touches are finalised. You may have already included these details in your plans, but there's nothing like walking through the house and seeing it for yourself. If you decide to move a power point by 300mm or a meter, it's a simple task that will only take the electrician five minutes to do if you do it now. The longer you wait, the bigger that job will get.

As you're deciding on the position of your power points, you should also finalise where all of your cabinets, benches, counter tops, vanities and wardrobes are going to go at the same time because these fittings and power go hand in hand.

ENERGY-EFFICIENT FITTINGS

According to the World Wide Fund for Nature, or WWF, the average Australian household emits around fourteen tonnes of greenhouse gases every year, making Australia one of the worst environmental offenders in the world. We can each do our part to reduce Australia's carbon footprint - and save a bundle on our energy bills at the same time - by using energy-efficient fittings. Here's what you can do.

Water taps

Install a low-flow or aerating tap and you could save as much as sixteen litres of water per minute compared to a less efficient tap.

Showerheads

A regular showerhead uses about fifteen to twenty five litres of water per minute – buy one with a three-star energy rating and that number drops to as little as six or seven litres a minute. Using less water means you heat less water, which translates into savings on your gas bill as well.

Toilets

There was a big push several years ago for Australians to replace their old single flush toilets with energy-efficient dual-flush models. It was a great idea because those single flush toilets can use up to twelve litres of water with just one flush – compare that number to the four litres (or less) a dual-flush toilet uses, and you can see why the new style is a better bet. It saves about 51 litres of water per person per day.

Reusing greywater

Another way to save water is by reusing your **greywater**. You can recycle your greywater by using it to water your lawns or flush your toilet.

Greywater is the water that comes from every water source in your house other than the toilet.

Appliances

Though technically not a fitting, a lot of people buy new dishwashers, washing machines, ovens and refrigerators when they build new or renovate their home. There are a lot of options these days for buying appliances that save on water, electricity and gas. Look for a minimum three-star rating – a five- or six-star rating is even better.

Lighting

Today's low-energy or energy-saving lights create the same amount of light but use a lot less wattage. Things like CFLs (those twisty fluorescent bulbs) may cost a bit more, but they only use about twenty per cent of a standard bulb and they can last anywhere from four to ten years – which is cheaper than buying new bulbs every year and creates less waste in general. Also consider using the new LED bulbs which are 80 per cent energy saving and last up to fifteen years.

WHERE TO FIND YOUR FITTINGS

If you haven't already done so, now is an excellent time to choose all of your fittings.

Hardware stores

If you're the sort who likes looking at a standing display, you might want to start at a major hardware store or home design store. These stores can provide you with individual elements or complete package kits. Kits can save you money, although the product often isn't as good as designing it yourself and there's a good chance you won't be able to install it properly.

Specialty shops

In addition to the big suppliers, there are innumerable speciality shops that deal specifically in kitchen and bathroom fittings. It's a matter of personal preference, but so much goes into installing a kitchen or a bathroom correctly that I would recommend you use a specialist company for these rooms, just to be sure you get it right.

Online

If you shop around and compare prices, you can save a lot of money. Probably the best way to find out what's available and at what price is by researching online.

Independent cabinetmaker

You can also hire an independent cabinetmaker who can design what you want and put it all together for you. They may be a bit more expensive, but a good cabinetmaker can design exactly what you want – which is perfect if you have an abnormal space or you have specific requirements that don't normally come in a standard kitchen or bathroom.

A Word to the Wise

Use your noggin

Before you get to the plastering stage, you need to make sure all the noggins are in. Noggins are the small pieces of timber that sit horizontally between the vertical studs. Noggins can hold up paintings, handrails for stairs,

towel rails, curtains, blinds, TV brackets – anywhere you need something solid behind your plaster.

This is another place where building your house yourself pays off because you can envision how the finished room will look. You can determine the best placement for those noggins to support the fittings you're going to want once you move in. You can't simply put a screw into a sheet of plaster – it has to have timber, steel or brick behind it to cope with the weight of whatever you're hanging on the wall. This is where you can take the most advantage of your noggins.

A Word to the Wise

Hang in there

If you or anyone working in your house is going to be up on a ladder, take extra care. I remember I was once working on a job on a first floor.

It was very early in my career, and like a lot of young blokes, I figured I was bulletproof. When I arrived on the job site one day, I discovered I was the first person there but I thought I'd go ahead and get started working anyway. It was raining that day, and the extension ladder we were using was resting on slippery tiles. I started climbing up the ladder and as I got halfway up, the bottom of the ladder started to slip out from underneath me.

I was able to grab on to the edge of the floor above me before the ladder slipped all the way out from under me.

For several minutes, though, all I could do was hang there about six metres above the tiles below me.

Luckily, one of the other guys arrived soon and was able to put the ladder back underneath me. If he hadn't got there when he did, I probably would have fallen and splattered myself on the ground. I was stretching the rules with safety and it wasn't smart.

CLEAN UP THE INTERIOR

At this point, you need to make sure the interior of the house is completely clean. Absolutely every single piece of rubbish should be removed from the job site and the floors should be thoroughly swept.

There will be a lot of mess still to come, so you're not finished yet but you need the house to be spotless at this stage to avoid interfering with the plastering.

PLASTERING

Before the final fit off, the plasterer will come. He'll start by simply hanging the plaster on the walls and ceilings. Once the sheets are up, again you should walk around the house and decide if you want to make any changes. If you do, those changes will need to happen before the plasterer starts putting the topping coats on to seal the plaster. Once coating starts, it's basically too late to make any further changes.

This tradesman is very important to the overall outcome of the interior of your house. I would want a plasterer to have a minimum of seven or eight years of experience, I would want to speak with other clients with whom the plasterer has worked and I would want to look at two or three previous jobs to make sure the work is up to scratch before contracting for this job.

If you have a bad plasterer, your house is going to look terrible – all of the joints are going to be visible or you might have curved walls. If the plastering isn't right and then you paint, that's where the faults in the plastering will become evident. If you have to have even one room replastered and repainted, it's going to cost you at least a $1000 for each wall or maybe even $2000.

In your kitchen, bathrooms and laundries, you'll have pipes and wires sticking out of the walls. In the bathroom, your carpenter will need to install bath frames. Then, working around the pipes and wires, your plasterer will line those walls and get those rooms ready for the tiler.

While your plasterer is working, there should be no one else in the house. That way your plasterer can get in and out as quickly as possible. Plastering for an average house from beginning to end will usually take a week including hanging the plaster, putting on the bottom coat, putting on the topcoat and then sanding.

Plastering is a big process and you may want to hire another professional to look at the work that has been done because you probably won't be able to see the imperfections whereas a carpenter or someone similar will be able to detects any faults.

CARPENTRY, PLUMBING AND ELECTRICAL

Once your plastering is complete, now it's time for the rest of the fittings to go in. This stage is still not the final fit out – some of these tradesmen may need to come back after the tiling is done.

Cabinets

Your cabinetmaker is usually the one to install your cabinetry such as cabinets, benches, bench tops, vanities and wardrobes – although kitchens are best left to a professional kitchen company.

If you get a kit, you can try to install it yourself or have your carpenter or another tradesman do it for you, but I wouldn't recommend using a kit.

Plumbing

Once the plasterboard is up, the plumber can bring all of the pipes through the walls and fit off the taps, baths, basins, toilets, sinks and troughs.

Electrical

Consult your electrician about when to install your power points, light fittings and light switches. Also, your electrician is the person to coordinate your air conditioning and/or heating provider to make sure those services are ready to turn on.

The electrician can do the fit-off either before or after painting – ask what he or she prefers and plan accordingly.

Telephone/Internet

Check with your service provider to schedule the best time for someone to come and place the connection points for your landline and/or Internet.

A Word to the Wise

Watch your equipment

It's a good idea to make sure your temporary power pole is working properly. I remember working on one job where the safety switch on the temporary power pole had a two-second delay. There were puddles everywhere and we had a few dodgy leads. Every time one of those leads fell into a puddle, we would get a zap. By the time the safety switch kicked in, it was too late – two seconds is more than enough time for a bolt of electricity to make it all through your body. Always check your leads and if something is in poor repair, fix it or replace it.

WATERPROOFING AND TILING

Wet areas like bathrooms and laundries require special attention – if they're not built properly, they can cause a lot of damage.

Waterproofing

Before any tiling can happen, all of your wet areas need to be waterproofed. The waterproof membrane creates a shell that traps all the water and keeps it from damaging your plaster, walls and subfloor.

Normally the tiler will do the waterproofing before doing the tiling, although in some situations the builder will do it. Either way, this job has to be certified by the person who does the work. You will not get your Certificate of Occupancy without a **waterproofing certificate**.

A **waterproofing certificate** is the document you get from your waterproofer and/or tiler (usually the same person) verifying that the area behind your tiles has been waterproofed.

You must make sure the person doing the job can certify the work once it's been done. If you don't get this certification, you'll have to pull it all apart and do it again.

What's the best way to do kitchen and bathroom renovations?

If you're doing a renovation, you can go through the process of designing and building a new kitchen or bathroom, but a better option is to hire a bathroom or kitchen specialist to do the whole lot for you. You'll have a lot of choices, so as always you'll want to do your homework.

Tiling

After the waterproofing is done, you're ready for tiles. The tiler comes in and puts the tiles on the walls, leaving holes around all exposed cables and pipes. Next comes the floor tiling in wet areas, so you may want to put down a tarp or some other protective covering on the floors in those spaces once those floor tiles are in place.

THE FINAL FIT OFF

At this stage, you're ready for your final fit off. Your electrician will finish installing power points, light switches, light fixtures and all other electrical fittings such as heated towel rails and power to all of your outdoor structures and lights.

The plumber will fit off tap ware, toilets, tubs and basins plus any plumbing necessary for swimming pools or other outdoor use. This is also the time to fit off the heating, air conditioning and solar systems and, if it hasn't already been done, your telephone/Internet connection points will go in now.

A carpenter will fit off vanities, internal doors, sliding doors, towel rails, mirrors and medicine cabinets plus **architraves, skirting boards** and door facings.

Architraves are the frames around your windows. Like door frames, they are usually made from wood.

Skirting boards – also called baseboards – are the strips of wooden boards that run along the bottoms of your interior walls.

Your wardrobes will also go in – you can have your carpenter put them in (either from a kit or custom made) or you can get a closet/wardrobe specialist in to fit off those areas.

These fittings can go in after the plastering and either before or after the tiling, depending on what type of product it is. At this stage, you can arrange to have more than one tradesman in at the same time, although you don't want to have too many people in the house working on top of each other. Make sure they're all spread throughout the house.

A Word to the Wise

Watch the workmanship

If a tradesman comes on site to do a job and realises the previous job hasn't been done properly then the time to fix the mistake is straight away. Don't let the problem linger and get forgotten about or built over. You don't want to pull your house apart to fix one little mistake that could have been repaired in ten minutes if you had caught it early.

HOW CAN I HELP?

Skilled professionals should do all electrical, plumbing, heating and cooling work – you should never get involved in those areas unless you are qualified. As always, though, there are jobs you can do, and at this stage the main one is cleaning.

Plasterers make a huge mess. For a whole house, you will fill at least a six-cubic metre bin full of off cuts. You'll find there will be dust all over the house, and if you haven't included clean up in your plasterer's contract, you'll need to constantly sweep up.

For people who work during the day, this is one of those things you can do in the evenings after the tradesmen have gone so that the site is ready for the workers the next day. You may even be able to get your children to help as long as the site is safe for them.

Fittings like cabinetry and vanities often come delivered in cardboard boxes. If you can't get the delivery people to take that packaging away, you can break down those boxes and haul them to the local tip for recycling and carry other plastic packaging to the bin.

EXTERIOR DESIGN AND LANDSCAPING

As the final fit off happens inside, it's time to finish off things outside, including driveways, paths and landscaping. Before doing your garden, you want to go ahead and finish putting in any other structures that will go up outside, including sheds, decks, pergolas, swimming pools and fences.

Once you have all these structures in place, design your garden to accommodate those structures. If you do your yard first, there's a good chance the tradesmen will wreck your landscaping just by walking through or by dropping tools on your plants.

Exterior design

Decks

When building a deck, you'll go through the same process as with a subfloor. Decks are usually built over stumps – either concrete stumps or timber stumps – or you may have a concrete slab with battens. I prefer to use a carpenter to do my decks because basically its carpentry, so a landscaper may not be qualified to do that job. You also have to keep in mind that all decks have to be built to code, so check with your local council. Your stump holes will have to be inspected to make sure they're at the right depth according to your soil report. Once that's been approved, you can install your stumps and complete the same process of putting in bearers and joists then laying your deck.

A Word to the Wise

Move it

There will be times when you really can't help so it might be best for you to get out of the way, especially if you're just doing a small renovation. Tradesmen don't want to feel like you're inspecting every second of their work, so if there's nothing you can do, just stand aside.

Fences

Before putting up a fence along the edge of your property, make sure you double check your land surveyor's report.

The last thing you want is to build a fence and then find out you're on your neighbour's property – if that happens, you can count on having to tear down the whole lot and start over.

Pergolas

A lot of Australian homes include some sort of covered indoor/outdoor space like a sunroom, covered deck or freestanding gazebo. You should have included these structures in the original design of the house – if you're adding it as an afterthought, make sure the foundation and other structural elements of your outdoor space won't compromise the construction of your main home.

Pavement

This is also the time for pouring pavement such as driveways and walkways. You might also have a patio area made of pavement, bricks or tiles rather than decking. You can have either a landscaper or a carpenter put in the pavement, but I would recommend a carpenter.

Swimming pools

If you included a swimming pool in your original design, you should have done the excavating and poured the concrete shell back during the foundation stage. At this point, all you need to do is have your tiling and plumbing fitted off. If you've only just now decided to add a pool, you may have trouble getting the excavating machinery into the proper position. There are options, though, like using a smaller excavator or putting in an above-ground pool, so ask your foreman or architect for ideas.

Outdoor power, water and gas

You should have had your electrician run cables to all of the exterior structures and features back before the cladding stage. If you didn't do that, you'll need to have your electrician go back now and wire all of those sources that require electricity to the main switchboard of your house. The same is true for any plumbing or gas lines that need to go in for which you will need a plumber. Do it now - better late than never.

Landscaping

Especially with a fresh plot of land, your yard space is likely to be quite muddy, and until you have grass or some other type of landscaping, you'll keep bringing mud into your house. For this reason, it makes the most sense to finish at least the core of your landscaping before you finish your floors. You don't need to plant every flower, but even basic landscaping will keep your yard from being a quagmire.

It's also a good idea to make sure the landscapers don't come into the house – if they do, make sure they are very careful to remove their shoes so they don't track mud into the house. Muddy boots can stain a floor and it may cost you thousands to get it cleaned or replaced.

A Word to the Wise

Give a little

One reason I love my job is that it's usually lots of fun. All sorts of practical jokes happen on site, everything from tampering with food and drinks to hiding tools or extinguishing a cigarette in someone's half drunk can of Coke.

You don't want to be a pushover, and you want your crew to take you seriously. At the same time, it's good having fun on site and you can help by getting together with the crew for a few beers after work at the end of the week. It keeps morale high and everyone happy.

As an owner builder, it doesn't hurt to build a little rapport with your employees by buying the occasional pizza or bringing coffee to everyone. It can make a big difference in what you get back in return. Doing those little things – we call them 'one per centers' – can really motivate your team to give you their best.

If you're happy with the work your employees are doing, let them know. There's nothing worse, from my point of view being a foreman, than doing a job and thinking I'm doing a good job and not hearing that from the client. We all want a pat on the back if we're doing a good job, so if praise is warranted, give it.

Sometimes tradesmen will take advantage or not appreciate your efforts, but most of the time, having that good relationship means those tradesmen will work harder for you than they might otherwise.

INSPECTING YOUR OUTDOOR STRUCTURES

Before you can move into your house, all of your structures are going to need to be checked out by your building inspector. It's always a good idea to include decks, pergolas and such in your original plans because eventually they have to be inspected by a building inspector. If you haven't done that and you build a pergola or a deck illegally, the council could come and make you pull it down. The same goes for a fence.

With certain renovations, you can sometimes skip the inspection of outdoor structures – especially if you already have a Certificate of Occupancy – but it's always a good idea to have these structures inspected anyway, and if you're not 100 per cent certain, ask. You may be able to arrange to have the inspector do your final frame inspection for your house and the stump inspection for your deck at the same time so that it doesn't cost any more money.

PAINTING

On *The Block*, everyone does their own painting, but it's still not the sort of job you should do unless you feel completely confident that you can do it perfectly. Painting is a very important job because everyone who walks into your house will see it.

I would recommend getting a professional to undertake this task, especially if you've got timberwork that's going to be stained and lacquered. It's too easy to ruin this step, and it isn't cheap. The average newly-built house will cost $10,000 or $15,000 to paint, and that's if you get it right the first time. A rescue paint job – one that corrects the mistakes you or another painter has made – will likely cost you even more.

Having said that, painting is one of the last jobs you'll need to have done, and if things have gone wrong on your build, you may need to make up your budget by doing the painting yourself. If you insist on doing your own painting or you really can't afford to pay a professional, you should adhere to a few key rules.

Tips for DIY painting

I've said throughout this book that when it comes to an area that isn't your specialty, you should always ask for advice from the experts – I'm the last person to tell you how to paint, so I called Barry Davies. A long time master painter and decorator, Barry is the owner of Barry Davies

Painting in Melbourne, plus he does all the painting on *The Block* that isn't done by the contestants. These are his top tips for painting the interior of your house yourself.

Start with the right tools

If you want your paint job to look professional, you really should use professional tools. For brushes and rollers, I use a brand called Durol. Monarch, Express Rollers and Oldfield's products are also very good.

You'll need rollers for all of your ceilings, walls and doors plus brushes for cutting in around the corners, windows, doors, frames and skirtings. You'll also need a smaller brush to do the detail work of your woodwork, including your door frames, architraves and skirting boards.

You'll need a couple of different rollers to do the job, each with a different **nap**. You'll need a roller with at least 18mm nap for your ceilings and one with a little bit less, maybe 12mm, for your walls. There are rollers that pump paint directly into the roller but I never use them and I don't recommend them.

The **nap** is the thickness of the material from which a paint roller has been made – the longer the nap, the more paint the roller will hold.

In addition to brushes and rollers, you'll probably want to pick up a few drop clothes, trays and possibly painter's masking tape. If you buy your paint from a proper painting store, they can probably tell you the best types of brushes to use in applying it.

A lot of the major hardware stores won't stock the proper trade brushes and rollers that professional painters use – for those you'll have to go to a paint shop. The good quality brushes are worth the money, though.

Prepare your surfaces

If you're doing a renovation, you'll need to be very careful to cover any areas that you don't want to get paint on. Painting can be very messy work, even for the professionals, so make sure any floors, carpets and furniture are well protected. If you're painting woodwork, it has to be filled correctly. There should be no gaps around the walls, the frames and the woodwork. We use a wood filler such as Polyfilla and then sand it back. There's another product called Uni Filler which we can lay in a very fine coat over the filler, if necessary – then we can sand it back to create a perfectly smooth surface.

If you're painting woodwork that's been glossed before, give it a good rub down with 120 or 180 grit sandpaper to make sure the surface is as smooth as possible.

A sealer will have to be applied to most newly-plastered walls. Sometimes these surfaces will have been pre-primed, but if they haven't, you'll need to seal them. You wouldn't normally put sealer on a ceiling – the only time I would put a sealer on the ceiling is in bathrooms and other wet areas. The biggest mistake I see in other people's paint jobs is that the filling and sanding hasn't been done properly. We have a sanding block that attaches to a pole to give the walls a bit of a light rub down and remove any little lumps or grit. In a new house, there's likely to be a bit of plaster dust left over, so you want to create as smooth a surface as you can before you start painting. In every house we do, we sand down the walls.

Undercoat your woodwork

Your door frames, architraves and skirting boards are probably made of wood, and this wood will hold paint differently than plaster. For woodwork, you can use acrylic but I recommend you use an oil-based undercoat plus an

enamel or satin gloss. These days, most people paint their woodwork to match the walls or maybe slightly lighter or darker for contrast. If you're painting your woodwork, start there – you should have already filled and sanded it.

Now you're ready for undercoat. We put a product called Penetrol in our undercoats and our glosses to help the paint flow out more easily when we're applying it. Mix a little bit of Penetrol or even a touch of gloss in your undercoat and then start brushing it on. Using smooth up and down strokes with a brush, paint all of the woodwork and leave it to dry for 24 hours.

Paint your ceilings

Most of the time, your ceiling will have been plastered, and you'll want to use a flat acrylic for painting plaster. The only time you won't use flat acrylic on your ceilings (and walls) is in a bathroom or wet area. In those rooms, you'll use a low sheen.

You'll probably find that water-soluble acrylics are easier to use than the oil-based enamel because acrylics go on smoothly, there's no smell, they dry quickly and they're easy to clean up. In a brand new home, you may be able to use a paint sprayer but I wouldn't recommend using a sprayer in an established home as it can make a huge mess. For a normal house ceiling, stand on a ladder for painting the corners and then attach an aluminium pole to the handle of your roller. You can get adjustable poles that extend out as far as 2.4m.

There are different methods, but the easiest is to start at one end and make regular strokes up and down along half the room then go back and do the other half.

One big pitfall is getting too much paint on one side of the roller and not enough on the other. To avoid that mistake, use a nice big 300mm wide tray, pour about four litres in the bottom and dip the roller into the tray. Work the roller around the tray so that you have the same amount of paint on all sides. Then lift it out and apply it.

Be careful not to get too much paint on your roller or you'll end up dripping it all over yourself. Also, if you don't apply the paint in an even coat, you'll end up with heavy **stipple**. This happens when the roller leaves behind a texture in the paint creating an uneven look – some very talented painters can apply a stippled look evenly to create a textured effect on the entire wall, but if you do it by accident, it will be an obvious flaw in your painting that you don't want.

Stipple is when a paint roller leaves behind a texture in the paint creating an uneven look.

For ceilings, we usually use two coats of a good quality flat acrylic. If you want to spend a little more money, you can put a sealer on the ceiling but I usually don't because it's not really necessary.

Drying time depends on what type of paint you use, although acrylics usually need four hours or less to dry between coats. Most paints come with guidelines for dry times printed on the side of the can plus the best type of roller or brush to use.

Paint your walls

Your walls have also been plastered, so again you'll use a flat acrylic for your walls just like you did with your ceilings. You can use the same roller for applying the paint to your walls as you used to apply the sealer, but you must roll it out carefully.

For the finishing coat on walls, you'll want to use your 12mm roller. If you use too fine a roller, you won't be able to get enough paint on and you won't get enough coverage. Use too much paint, though, and you run the risk of creating that unsightly stipple.

When painting walls, I like to start on the left-hand side of the room and work across (which side you start on – left or right – just depends on your preference). Using even up and down strokes, I move across the room, first covering just the top half of the wall from the ceiling to about the midpoint. Then I go back and do the bottom half, working my way down from the midpoint to the skirting boards.

Your walls will need a minimum of three coats of paint including sealer. You can put a lot of paint onto the roller but you've got to then roll it out onto the wall to spread it evenly. The trick is to get just the right amount of paint on the roller – not too dry and not too full. Also, don't go too fast because if you do, the roller will spin and the paint will splatter.

If you're renovating, existing walls only need two coats. If there are any water or smoke stains on your existing surfaces, you will need to apply a stain blocker to those affected areas before applying a final coat.

Repaint your woodwork

You will have already put on one coat of enamel on all of your woodwork. Now that your ceilings and walls are finished, you can go back and give your woodwork one final coat of enamel or satin gloss. You're using different types of paint – maybe even different colours – so you need to be able to cut in neatly around the woodwork so that you don't get the enamel on your walls. To cut in those areas, you can use an angled brush but professional painters usually use oval ones.

You can use masking tape to make sure you get a nice clean line, but I don't like to use it because sometimes it pulls the paint off the walls. A lot of people use the tape, but I've seen disasters where they've pulled the tape off and the paint's come off with it. If you are going to use tape, be sure to pull it off straight away – within 24 hours, at the most.

Touching up or repainting

If you're painting your new house for the first time, you'll need three coats for your walls, which can take several days. If you're just getting the interior of your house repainted – perhaps to sell or just to spruce it up – you're looking at two or three days, especially if there's furniture and carpets in the house. We'll normally go around the house and touch up any marks or stains and then apply one coat over everything. In most cases, that's enough to freshen up the paintwork.

Painting is a big job. With a brand new home, you're looking at about a week for painting and you should estimate $7000 to $10,000 depending on the size of your house and the types of paint you use.

Some painters work on a rate of about $25 per square metre, which is pretty poor. At those rates, they'll want to get in and get out quickly so they'll probably spray everything. It's a quick and easy method but there's no detail work involved and all of your woodwork will be the same colour – and coated in the same type of paint – as your walls. Other painters will charge $45 to $50 per square metre, but they will do a better, more detailed job for the price.

A Word to the Wise

An exception to the rule

We sometimes use flat enamel on ceilings rather than flat acrylic but only in very old homes where the ceilings are deteriorating a bit. If there's a problem inside the plaster, flat enamel can seal back any stains or other flaws in the ceilings but that's rare these days.

FLOORS

Now that everything else is done, it's time for floors. I'd recommend leaving your floor coverings for last because if you do it any earlier than now, you or your tradesmen will keep tracking mud into the house and you may ruin your floor coverings. Here are some options to think about.

Carpets

Your carpet installers shouldn't come until the last few days, but they shouldn't need more than a day to install carpet throughout your whole house.

Tiles

You'll only need waterproofing in laundries and bathrooms (but not toilets). If you're putting in a tiled floor anywhere in your house in addition to your wet areas, you'll probably have all of it done at the same time. If you get your tiling done earlier in the build, don't worry – tiles are pretty hearty, so when it's time for tiling, just get the job done and then cover the floor with a protective layer of plastic or tarp.

Floating floors

Because they simply rest on the subfloor and snap together, floating floors will take no more than a day to install throughout your whole house.

Floorboards

It's possible to install your floorboards early on, but it's better to wait if you can. Real floorboards could take maybe four or five days to install for the whole house. Then you've got your lacquering stage, which should take another four or five days. Once that's finished, you should allow at least two days to dry.

It's imperative that if you're going to lacquer your floor, no one goes inside your house at all while that job is being done. It's best to just seal off the house and not let anyone inside. In addition to running the risk of ruining your finish, you'll also find that the house is going to have a lot of fumes and it's quite dangerous to be in the house while the lacquer is drying. In other words, stay out and make sure you don't have any trades lined up to come into your house while you're getting your floor lacquered.

THE CERTIFICATE OF OCCUPANCY

Before you can move in, you'll need to obtain a Certificate of Occupancy from your building inspector who will come back for one final check and either sign off on your job and give you the Certificate of Occupancy or else give you a list of items that need to be fixed. If there are any

problems, you'll want to get those items repaired straight away. Then your inspector can come back and check it out again.

TIME TO MOVE IN

By this stage, you've done it all. You've designed your home, documented everything, overseen the build and inspected each step of the way. You even have your Certificate of Occupancy. Now you can move in!

CHAPTER SUMMARY

- After framing and lock up, it's time for fitting off, which means installing all of the final touches that make a house livable.

- Start by planning your power and fittings. Think about how you want to place your furniture and decide where every fitting will go to accommodate your needs.

- In choosing fittings, you'd do well to look for energy efficient options including water saving taps and showerheads, duel flush toilets, greywater reuse systems, energy efficient appliances and low energy lighting.

- You may be able to save money when shopping for fittings. Try looking at hardware stores, specialty shops, online or hiring an independent cabinetmaker.

- Place your noggins so that they can help support all the appliances and fittings that will go in after your plaster is up.

- Before plastering, make sure the interior of the house is spotless.

- Your plasterer will now come to line your walls and ceiling with cement sheet plus several layers of topcoat. It is vital that your plasterer does a good job, so choose wisely for this important job and keep everyone else out of the way.

- After the plastering, you'll bring in your carpenter, plumber and electrician to put in the rest of the fittings.

- [ ] If you're renovating a kitchen or bathroom, I recommend using specialist companies to fit off these important rooms.

- [ ] With all the fittings in place, your tiler can come in to waterproof and tile your wet areas then give you a waterproofing certificate.

- [ ] If anyone catches a mistake at this point, make sure it gets fixed immediately.

- [ ] The main job you can do to help during the fit-off stage is cleaning and taking packaging materials to the bin or to be recycled.

- [ ] Now is the time to finish all exterior structures including decks, fences, pergolas, pavements, swimming pools and any other structures outside that may require power, water or gas. Make sure to have these structures inspected.

- [ ] With the outdoor structures done, you can landscape – just keep any tradesmen with muddy boots outside the house.

- [ ] While the landscaping is happening outside, you can get the painting done inside. It really should be done by a professional, but if you must do it yourself, be sure to follow the rules as outlined by master painter Barry Davies.

- [ ] The very last job will be laying your final floor covering, including carpets, tiles, floating floors and/or floorboards.

- [ ] It's time for your final inspection – if you pass, your inspector will give you the Certificate of Occupancy and you can move in! If you don't pass, get any problems rectified and then reinspected.

WHAT COMES NEXT

You've done it! You've considered every angle, organised all your documents, worked with your architect, engineer and foreman, got sign-off by your building inspector – basically, you've designed your house and built it from the ground up. Two last things will complete the picture: adding your special touches to the interior and making sure that your property stays in good condition.

CONCLUSION

INTERIOR DESIGN, MAINTENANCE AND REFLECTING ON A JOB WELL DONE

You've got your Certificate of Occupancy and you've moved into your masterpiece. But wait, there are still some things to think about: decorating the interior and making sure your house stays in good order.

INTERIOR DESIGN

Much of what you see on *The Block* is about the interior decorating of the homes. When you first started looking at ideas for designing your home, you probably noticed the interior design first. Some people take years to decorate their houses, and it's a rare home that doesn't undergo several changes in style over the years.

If you know what you want in the way of feature walls, you can have your painters apply the colours based on your direction. If you don't have a colour scheme in mind before you start building, that's okay too. You can always decide later. In fact, unless you're very sure about your interior design ideas, you may want to start with antique white on all of your walls and make changes after you've lived in your house for awhile. That's what I did.

As you did with the overall design of your house, you'll want to do your research when it comes to interior design - look at magazines, look online or look at displays. Then you can set yourself a budget and choose for yourself, or hire someone else to decorate your house for you.

If you like the look of the homes on *The Block*, you should check out the tips in *Design Your Home*, a book on interior design written by *Block* judge Shaynna Blaze. She is much more qualified than I am to give you hints on interior design - and as you know by now, I'm all about asking the experts.

MAINTENANCE

You've built your house, so now all you need to do is take care of it. Home maintenance isn't scary, but if you don't do it, those little problems will quickly turn into big, expensive ones.

A little bit of regular home maintenance will make your house more comfortable and will save you heaps down the line. As with everything, the maintenance you need to do depends on your house, but these are just a few guidelines.

Once a week

Here are some maintenance jobs to do each week.

Cleaning

No one likes housework, but weekly cleaning is a good way for you to run your eye over all the corners of your house and make sure everything is running as it should. Look for mould, ants, leaks, dripping faucets or other problems that will just keep getting worse.

Gardening

Some shrubbery and plants require regular pruning, and no one likes an unsightly overgrown yard. Stay on top of your yard work and it won't grow out of proportion. Especially keep an eye on climbing vines - if you don't trim them regularly, they can get into your guttering and pipes, creating huge problems.

A Word to the Wise

Don't be a drip

Make sure your pipes are draining properly and your taps turn off easily and completely. Leaky pipes will never get better on their own, so don't let a little drip turn into a big one. Every drip costs you money and wastes our resources. If you do get a drip, fix it or hire someone who can.

Once a month

Some jobs should be done every month.

Clean your disposal

If you have a disposal in your sink, it needs regular cleaning - otherwise it may clog and start to smell. It's just a simple matter of removing any large bits of debris and running a sinkful of warm, soapy water through it. Just be sure you've switched off the power before reaching into the device.

Check your HVAC filters

HVAC systems have filters for capturing dust and other impurities.

HVAC stands for heating, ventilation and air conditioning.

When the filters get too dirty, the system doesn't work as well. Blocked filters may also present a danger, so check them regularly.

Clean your oven range filter

Skip this step and you could end up with a fire on your hands. Most oven range filters slip out easily so you can wash them with hot soapy water in the sink or dishwasher.

Clean dishwasher filters

To make sure your dishwasher works as well as when you first bought it, clean out the filters at least once a month.

Inspect your fire extinguisher

Hopefully you'll never need it, but if you do, you want to make sure your fire extinguisher is ready for use. Most regulation extinguishers come with instructions and dates of use printed on the side, so make sure yours are up to date.

Once a quarter

Every three months, it's a good idea to do the following checks.

Check your smoke and/or carbon monoxide detectors

Make sure these devices have fresh batteries and are in top working order. They could save your life.

Test your garage door

If you have an automatic garage door opener, it should come with a safety device that makes the door go back up in the event that something or someone gets trapped underneath the door. Double check that this feature is working properly, especially if you have small kids around.

Once a year

Yearly checks are minimal but worth the effort.

Check your trees

Make sure your garden isn't ruining and rotting your house. Keep an eye on your roots and your root systems. If a root is going underneath your foundation, it may crack your slab. It can also affect stumps, so keep an eye on all your roots, especially those of big trees. Also, look to see that large branches aren't interfering with power lines.

Clean out the gutters

Your gutters must stay clear – if they don't, your down pipes can get blocked causing water to overflow into or onto your walls, ruining your plaster and causing a lot of damage. Also, if you're using water tanks, blocked gutters will reduce the amount of rain that makes it into your tank, giving you less water to reuse.

Check your timber

Timber is a natural product and without protection, it will either rot or fade. If you have timber on the exterior of your house, you should give it a good looking over once a year and make repairs or repaint as necessary.

Check caulking and grout

Used for sealing your sinks, tubs, showers and tiles, both **caulking** and **grout** can deteriorate over time. Look for small problems and get them repaired before they turn into big ones – replacing all of your caulk can be very expensive so keep an eye on it.

Caulk is a thick, sticky sealant that dries to a rubbery finish. It is used to waterproof kitchen and bathroom fittings.

Grout is the mortar or paste between tiles.

Once every two years

Biannual checks are equally important as other checks as these can ensure that your home doesn't become toxic.

Get heating serviced

Heaters need to be serviced regularly, especially gas heaters because they can be toxic. Every year we have a few deaths in Australia caused by carbon monoxide poisoning, which can result from a poorly-maintained gas heater. Have either a plumber or a gas fitter look over all of your gas appliances - such as a gas water heater - and make sure you don't have any toxic leaks.

Get A/C serviced

All appliances need regular maintenance, and your air conditioning or evaporative cooling is no different. Parts sometimes break down, and you may not know it's happening, so once every year or two, have someone come to make sure yours is running smoothly.

Once every ten years

Here are a couple of big-picture maintenance items that shouldn't need your attention until the 10-year mark or thereabouts.

Repaint the exterior

Over time, your paint will fade and crack, especially if your house is clad in weatherboards or you have timber fascia boards. You'll need to paint all exterior woodwork every seven to ten years. Don't neglect the painting – if you do, you'll end up needing to replace those weatherboards or fascia boards, which costs a lot more than just painting. Also, if your exterior timber has cracks or holes, water will get in and ruin your walls. If you take good care of your timber, though, it can last up to 100 years.

Check out your interior paint

How often you need to repaint the interior of your house will depend at least in part on who lives there. If your house was painted properly the first time and you don't have children or pets, you may be able to go 20 years or more before repainting your interior. If you share you home with kids or animals, that job may need to be done more often. Just keep an eye on the paint inside your house and consider the best time to spruce it up.

Respray your roof

If you have tile roofing, you should look at getting it resprayed. Sun, wind and rain can weather a tile roof, making it less durable than it should be.

There tends to be a lot less maintenance these days than there used to be. A lot of modern materials – like Colorbond steel sheet roofing and enclosed powder coated guttering – don't need as much attention as older materials. When it comes time to do that essential home maintenance, consider replacing older, less reliable materials with more efficient low-maintenance materials.

REFLECTING ON A JOB WELL DONE

Building can be very gratifying work. Most of the work I've done throughout my career has been on high-end homes - I've built houses in some of the most well-known streets around Melbourne such as Albany Road in Toorak and Shakespeare Grove in Hawthorne.

I'm also really proud of the work I did as the foreman on the renovation of The Cosmopolitan Hotel in St Kilda. On that job, we updated the entire structure and turned 150 shabby rooms into 400 spectacular ones - we won The Best New Building in the City of Port Philip for that project.

And of course, I've overseen some fantastic work on *The Block*. I've seen around thirty houses and apartments built on the show, and we've had pretty much zero negative feedback, either from our competitors or the people who have bought the finished products.

Through it all, though, I think I'm still probably most proud of Polly and Waz. Even though they had zero experience or building skills, they were able to listen, learn and work hard to create a really impressive home. There's no reason why you can't do the same.

It's pretty cool to drive down the road and see a house I've built. It just gives me a real sense of pride. Now imagine if you could get that feeling every time you came home at the end of the day. If you're like me, you'll absolutely love the sense of accomplishment you'll feel when your house is finished and you're ready to start living in it.

A couple of things to remember:

- You may love your new home and plan to live there for the rest of your life - or you may look at this success and decide to sell your house and do it all over again. Depending on where you live, you probably have to live in your house for a number of years before you can sell it and guarantee the work for even longer. Check the rules for your area. As the owner builder, you should know your new house upside down and inside out. If you decide in the future that you want to make alterations or renovate, you know exactly how that house is put together.

- You also know every stick, every brick and every nail that went into that house, so you know what needs to be done to maintain it. Keep up with the regular maintenance and your house can take care of you for a long time.

- There's also a substantial financial reward. Remember how your lender wouldn't lend you the total value of the finished product? If you did your job properly as an owner builder, there's every chance you've built a home that's worth more than what you owe – perhaps a lot more. And if you've done renovations, you're likely to have increased the value of your home, making it into a real investment.

It may also be that your finances wouldn't allow you to build every single feature you had hoped to include or incorporate all of the materials you really wanted – this time. If so, there's a great chance you can sell your house in a few years' time and make enough in profits to do it all over again – this time with the money to build exactly what your heart desires.

There's no doubt that building a house is hard work. Volume and other independent builders will charge you that twenty or thirty per cent premium for good reason.

If you decide to take on that role yourself, you must be prepared to make a big investment of time and energy.

But if you're prepared to stick to your budget and your schedule for the relatively short amount of time it takes to work through the construction process, you're likely to come out ahead for decades to come. More than the financial value of your home, though, you have the most valuable thing – the perfect place to live. I'm rapt that I built my own house myself. The feeling of 'wow I built this' is amazing.

My house isn't the biggest or grandest structure I've ever been involved in building – but it's mine. It's where I go home every day, where my children are growing up and where I spend time with my wife and friends. It's my sanctuary, a place where I can work and rest and escape from everything – and it is exactly what I want.

With the right personality and attitude, plus a willingness to stick with the job, do your research, stay organised and work hard, you too can design the home of your dreams and turn it into a reality while saving heaps of money at the same time. Do it right and you'll be glad you did.

SUMMARY

- Interior design for your home may take years and you may change the design from time to time.

- There are endless options in terms of interior decorating. Check out the guidelines in Shaynna Blaze's book *Design Your Home*.

- Weekly maintenance includes house cleaning and gardening.

- Monthly clean filters for HUAC, oven range and dishwasher to keep things working properly.

- Each quarter, check smoke and carbon monoxide detectors to keep your home safe.

- Yearly, check trees, clean out gutters and check timber, caulking and grout.

- Every two years, have heating, gas and A/C serviced.

- Every ten years, consider repainting and respraying your roof.

- Reflect on a job well done and enjoy!

- You've now done the lot, right down to the smallest details of interior design and making sure your property will stay in good condition as you enjoy the years in your new home. What's left to do? Not much – except for maybe pat yourself on the back. Well done!

ACKNOWLEDGEMENTS

It takes a whole team of people to build a house, and the same is true for writing a book. I never could have written this book if it hadn't been for a lot of help from a lot of people.

First of all, I'd like to thank Diane Ward and Fiona Schultz from New Holland who believed in this project and helped make it happen plus the design and editorial teams who made the final product look fantastic. Also a big thanks to Laura Fulton for helping me put this book together. She's a clever girl, and together I think she and I have come up with a great result. I couldn't have done it on my own.

I'd also like to thank my colleagues from *The Block*, especially Barry Davies for providing a comprehensive guide to painting and Shaynna Blaze for writing a book that covers the one subject I know the least about, interior decorating.

Another guy who's been instrumental in my success on *The Block* is Julian Cress, the Executive Producer and one of the show's creators. He's an amazing and clever man and I've learned a huge amount from him, especially in terms of how to stay calm. He's taught me that every problem has a solution and how to keep my cool in the toughest situations. Of course, I never would have been chosen for *The Block* if it hadn't been for the support of a couple of guys who helped me develop in my trade. Over twenty years ago, my great friend Shane Smith helped me get into the field of building high-end houses, which is where I really wanted to be.

Through him, I also met Andrew Gorman, a project manager who showed a lot of faith in me early on in my career. He's given me the opportunity to work on some amazing projects over the years, and he's taught me a lot about carpentry, management and every other aspect of my trade. I owe a lot of my success in my career to Andrew Gorman.

Most importantly, I couldn't have done it without the love and support of my family. My wife has been hugely supportive of me, especially during the last four years that I've been with *The Block*. My work takes me away from home a lot, and she's been able to really take over on the home front, filling in for me in those 'dad' roles I'm not always there to do. She's always patient and obliging and I wouldn't be here today if it hadn't been for her support.

Thanks again to everyone!

Keith

GLOSSARY OF TERMS

Throughout this book, I've used a lot of technical building terms, some more complicated than others. Here's a brief recap of all the terms you've seen, just in case.

Ant caps, usually made from galvanised sheet metal, are designed to keep termites from getting into your subfloor.

An **architect** designs your entire environment, including all indoor and outdoor spaces.

Architraves are the frames around your windows. Like door frames, they are usually made from wood.

Battens are long flat strips of timber or metals. In a floor, they attach to the foundation to create a surface on top of which to lay floorboards. You'll also have battens in your roof – they create the surface on which your tiles or steel sheet will sit.

Bearers are part of the subfloor – these lengths of hardwood sit directly on top of your stumps to create a solid base for your house.

Blueprints are the precise technical drawings an architect will have a draftsman draw up – your tradesmen will refer to these drawings in building each step of your house.

Brick piers are a type of foundation made from a series of pillars made from bricks that rest on top of a concrete pad.

A **builder's pole** is a temporary power pole that is separate from your existing power, supports a big power box to run all of your leads out of and has a safety switch so it eliminates the chance of getting electrocuted.

The **building inspector** – sometimes called the building surveyor – will check your project at various stages throughout the build to insure that all of the work has been done properly. When you pass the final check, the inspector will give you the Certificate of Occupancy.

Caulk is a thick, sticky sealant that dries to a rubbery finish. It is used to waterproof kitchen and bathroom fittings.

The **Certificate of Occupancy** is the document that proves your home has passed all stages of inspection – without it, you won't be able to insure your home.

Cladding is the façade of your building, the external layer that will be visible on the outside of the house and covers the internal structural frame of the house.

Construction insurance covers your property in the event of vandalism, theft, inclement weather, fire or other catastrophes – it can even cover tools and personal injury.

C**onstruction loan** – your lender would calculate the total amount your project is going to cost then divide that figure into four or five separate payments called progress draws which would be transferred directly to your builder in chunks as the project progresses.

Contractors – also called tradesmen – are the skilled workers who provide the labour and/or materials necessary to build your house.

The ones you're most likely to need are concreters, carpenters, plumbers, electricians, plasterers and painters.

A **credit risk** is someone who has failed to pay his or her bills on time in the past. A lender won't want to loan you money if you're a credit risk because they have reason to believe – based on your previous borrowing history – that you may not pay them back.

Domestic building insurance will cover the person who buys your home in the event that you die or go bankrupt and your house is found to be incomplete or defective.

A **draftsman** creates precise technical drawings of your house called blueprints usually based on an architect's designs.

The **eaves** of the house are the bottom bit of the roof that hangs over the tops of the walls. They allow water to run away from your house rather than down the walls.

The **engineer** is the person who works out the science and maths involved in turning your designs into a reality.

The **envelope** is the three dimensional zone into which your entire property needs to fit, including length, width and height.

Excavation means digging out the foundation for your house, usually with a large digging machine like a Bobcat.

A **fall** is the tiny amount of slope in the floor of a wet room that allows water to drain.

Fascia boards are the long, straight boards that run along the lower edge of the roof – they usually support the bottom row of your roof tiles and your gutters are usually attached to them.

Fitting off (or the fit-off stage) is the final work that makes a house livable. It is the last stage of the building project and involves the last of the carpentry, electrical work, plumbing work, painting and flooring.

Flashings are the seals that go around your windows and external doors. They are essential for protecting your home against the weather.

A **floating floor**, which is usually made of vinyl or laminate, sits directly on top of your subfloor or slab and fits together in pieces.

Footings are the structural base of your house, what prevents your house from shifting or moving. Your foundation will sit on top of these footings.

A **foreman** is in charge of overseeing your build and liaising between you and the rest of your employees.

The **frame** of your house is the interior structure of your walls. Usually the frame is covered with some other material.

A **granny flat** is usually a freestanding apartment or studio built on your property. It can come with more or fewer amenities, such as bathrooms and kitchenettes, depending on your needs.

Greywater is the water that comes from every water source in your house other than the toilet.

Grout is the mortar or paste between tiles.

HVAC stands for heating, ventilation and air conditioning.

Income protection insurance will provide you with money if something happens and you stop earning your regular income.

Joists are lengths of hardwood that lay perpendicular on top of the bearers directly beneath them. The joists are closer together than the bearers and create a sturdy grid on top of which you'll lay your floor covering.

A **land surveyor** measures the exact boundaries of your property to a high degree of accuracy.

The **lock-up stage** is when the subfloor, all the exterior walls, the roof, the exterior windows, doors and cladding are put in place. In other words, the build has progressed to the stage where you can lock out prowlers and the weather.

LVL stands for laminated veneer lumber. LVLs are made from two pieces of 90x45mm wood that have been laminated together to create a thickness of 90x90mm. They're great protection against termites.

A **margin** – also called a premium – is the amount of money a builder will charge you in addition to the cost of labour and materials. Basically, it's the amount you pay that your builder takes as profit.

The **nap** is the thickness of the material from which a paint roller has been made – the longer the nap, the more paint the roller will hold.

Noggins are the small pieces of timber that sit horizontally between the vertical studs in the frame of a wall. See also 'studs'.

According to the Victorian government, an **owner builder** is someone who takes on most of the responsibility for domestic building work carried out on his or her land.

With an **owner builder mortgage**, your lender calculates the percentage of your build you can borrow based on quotes from your tradesmen and periodically transfers the progress draws directly to you.

PPE stands for personal protective equipment. It refers to clothing designed to keep you safe on a jobsite such as hard hats, gloves, safety glasses, safety boots, ear protection and high visibility vests.

Prefab – short for prefabricated – refers to a method of framing whereby the frame of the walls or roof are designed by AutoCAD and then built and assembled in a factory.

A **premium** – also called a margin – is the amount a builder will charge you in addition to the cost of labour and materials. Basically, it's the amount you pay that the builder takes as profit.

Progress draws are the payments your lender makes either to you or your builder as your project progresses. By the end of the build, your lender will have disbursed the entire amount of the loan.

Public liability insurance protects you if anything happens to the public as a result of your build.

Rendering means covering another type of material with plaster to create a smooth surface.

Sarking is insulation that goes on both roofs and some types of walls.

A **screed** is a short length of wood or aluminium the concreter will use to smooth out the concrete so that your slab will dry as a flat, level surface.

Skirting boards – also called baseboards – are the strips of wooden boards that run along the bottoms of your interior walls.

A **slab** is a type of foundation made from a single, thick layer of concrete that gets poured all at once to create a solid base on top of which to build your house.

A **solid brick** frame is composed of a brick frame and plus a second layer of brick veneer.

A **stick built** frame for a wall or roof is constructed piece by piece by a carpenter.

Stipple is when a paint roller leaves behind a texture in the paint creating an uneven look.

Structural mesh – also called trench mesh or rebar – is a grid of metal designed to give the concrete of your foundation a spine.

A **stud** is piece of timber that is used to create a frame. The long studs stand upright and generally run from floor to ceiling. A noggin is the small piece of timber that sits horizontally between the vertical studs. See also 'noggins'.

Stumps are a type of foundation made from a series of posts (usually made from concrete or hardwood) that have been embedded into the ground and set in place with concrete.

The **subfloor** is the layer of flooring that is built over a stump foundation, creating a flat, solid surface.

Theodolites are highly precise instruments that make exact horizontal and vertical measurements, mainly for the purpose of establishing that a surface is level.

Thermal mass refers to how well a material can absorb and hold heat – a slab generally has a lot of it while stumps and piers have less.

Tradesmen – also called contractors – are the skilled workers who provide the labour and/or materials necessary to build your house. The ones you're most likely to need are concreters, carpenters, plumbers, electricians, plasterers and painters.

A **trussed roof frame** is designed to create a grid of timber on top of which to lay your roofing materials – the truss is one of the supporting legs of the frame.

Underpinning refers to a foundation of stumps that are already in place.

The **White Card** is required for everyone who does construction work in Australia – it is the common term for General Safety Induction Training.

A **waterproofing certificate** is the document you get from your waterproofer and/or tiler (usually the same person) verifying that the area behind your tiles has been waterproofed.

Workman's insurance – also called WorkCover – protects the people who work for you in the event that one of your tradesmen gets hurt while working on your house. It covers benefits such as income replacement, medical treatment, rehab, legal costs and even lump sum compensation if the injury is serious.

Yellow tongue – also known as chipboard or particleboard – is the most common flooring material. It's not very attractive, but it's water resistant and usually pretty cheap.

APPENDIX

Master Schedule

Here's a sample of a fictional master schedule for building a typical family home, just to give you an idea. Notice that this schedule doesn't include weekends or other holidays - yours will. As you draw up your schedule, you can also include how much you'll need to pay for materials and labour on each day when those payments are due.

Day 1 Excavate for stump holes and garage slab	Day 2 ■ Stump hole inspection 8:00am ■ Pour concrete for stumps 9:00am ■ Put in steel for garage ■ Install stumps	Day 3 ■ Garage slab inspection 8:00am ■ Pour garage slab 9:00am ■ Finish garage slab by the end of the day	Day 4 Install bearers and joists	Day 5 ■ Finish installing bearers and joists ■ Dig trenches for sewer and storm water	Day 6 ■ Put up wall framing ■ Finish digging trenches for sewer and storm water ■ Dig trenches for outdoor electrical	Day 7 ■ Finish wall framing ■ Install sewer and storm water pipes ■ Install outdoor power cables in trenches
Day 8 Put up roof framing	Day 9 ■ Put up roof framing ■ Finish brackets and tie downs ■ Frame inspection	Day 10 ■ Straighten external walls ■ Install roof sheeting	Day 11 ■ Install external doors, windows ■ Install roof sheeting	Day 12 ■ Finish installing external doors, windows ■ Finish installing roof sheeting	Day 13 Install weatherboards	Day 14 Install weatherboards
Day 15 Install weatherboards	Day 16 Install weatherboards	Day 17 ■ Finish weatherboards ■ Straighten internal walls ■ Paint exterior	Day 18 ■ Rough in plumbing, electrical, A/C, heating ■ Paint exterior	Day 19 ■ Rough in plumbing, electrical, A/C, heating ■ Paint exterior	Day 20 ■ Install insulation ■ Install gyprock, cement sheet ■ Paint exterior	Day 21 ■ Finish gyprock, cement sheeting ■ Put in eaves and fascia

Day 22 ■ Apply base coat of plaster ■ Put in eaves and fascia	**Day 23** ■ Apply top coat of plaster ■ Install gutters	**Day 24** ■ Sand plaster ■ Install gutter	**Day 25** ■ Tiling ■ Doors, skirting, architraves	**Day 26** ■ Tiling ■ Doors, skirting, architraves	**Day 27** ■ Tiling ■ Doors, skirting, architraves	**Day 28** Tiling
Day 29 Tiling	**Day 30** Painting	**Day 31** Painting	**Day 32** Painting	**Day 33** Painting	**Day 34** Painting	**Day 35** Painting
Day 36 Painting	**Day 37** ■ Install kitchen ■ Fit off plumbing, electrical	**Day 38** ■ Install kitchen ■ Fit off plumbing, electrical	**Day 39** ■ Install kitchen ■ Install downpipes for gutters	**Day 40** ■ Install floorboards ■ Finish downpipes and water tank	**Day 41** Install floorboards	**Day 42** Install floorboards
Day 41 Install floorboards	**Day 41** Sand floorboards	**Day 41** Coat floorboards	**Day 41** Coat floorboards	**Day 41** Install floorboards	**Day 41** Install carpets	**Day 42** Put in landscaping, driveways and paths
Day 50 Put in landscaping, driveways and paths	**Day 51** Landscaping	**Day 52** Final inspection				

INDEX

Photo credits

Front cover image, prelim pages, pp. 2, 4, 6, 8, 318 and all images of Keith used with permission from Steel Building Systems Australia Pty Ltd

Pages 46, 54, 63, 66, 71, 75, 89, 93, 96, 120, 126, 139, 148, 157, 161, 168, 179, 184, 196, 200, 209, 216, 221, 232, 240, 245, 256, 266, 292, 313 – Shutterstock, www.shutterstock.com

Pages 80, 112, 120, 139, 248, 273, 280, 332 – freeimages.com

Page 20 mirofoto, www.sxc.hu/gallery/mirofoto

Page 27 Bill Twyman, bill.twyman@mac.com

Page 35 Maira Kouvara, freeimages.com

Page 42 Graham Briggs, freeimages.com

Page 83 Macin Smolinski, www.facebook.com/24youPhotography

Page 86 Barun Patro, freeimages.com

Page 99 Dirk Dratzenberg, freeimages.com

Page 100 Yucel Tellici, freeimages.com

Page 104 Barum Patro, www.barunpatro.com

Page 132 and backcover watermark with permission from Debra Williams, JacksonTeece Architects, www.jacksonteece.com

Page 207 Oliver Gruener, freeimages.com

Page 213 Mario Magallanes, freeimages.com

Page 285 Andreas Krappweis, photography@krappweis.com

Page 299 Krzysztof Szkurlatowski, www.12frames.eu

Page 304 Peter Hellebrand, freeimages.com